I am making no claims for the survival of the personality; I am not promising communication with those who have passed out of this life. I merely state that I am giving psychic investigators an apparatus which may help them in their work, just as optical experts have given the microscope to the medical world. And if this apparatus fails to reveal anything of exceptional interest, I am afraid that I shall have lost all faith in the survival of the personality as we know it in this existence.

Thomas Edison

Matters of fact well proved ought not to be denied because we cannot conceive how they can be performed. Nor is it a reasonable method of inference, first to presume the thing impossible and there to conclude that the fact cannot be proved. On the contrary, we should judge of the action by the evidence and not of the evidence by the measures of our fancies about the action.
Joseph Glanville, F.R.S.

The popularity of the paranormal, oddly enough, might even be grounds for encouragement. I think that the appetite for mystery, the enthusiasm for that which we do not understand, is healthy and to be fostered. It is the same appetite which drives the best of true science, and it is an appetite which true science is best qualified to satisfy.
Richard Dawkins

The photography of spontaneous ghosts is a chancy and unreliable business Photographs of ghosts may be the most valuable independent evidence yet obtainable, but the malicious production of so-called 'ghost photographs' is notoriously easy and, in some cases, the results are extremely convincing to anyone other than a person experienced in photography and fraud.
Peter Underwood

The Haunted Field Guide Series

STRANGE FREQUENCIES
A Practical Guide to Paranormal Technology
BY CRAIG TELESHA

- A Whitechapel Press Book from Dark Haven Entertainment -

© Copyright 2008 by Craig Telesha
All Rights Reserved, including the right to copy or reproduce this book, or portions thereof, in any form, without express permission from the author and publisher

Original Cover Artwork Designed by
© Copyright 2007 by Michael Schwab & Troy Taylor
Visit M & S Graphics at http://www.manyhorses.com/msgraphics.htm

Editing & Proofreading Services: Jill Hand

This Book is Published By:
Whitechapel Press
A Division of Dark Haven Entertainment, Inc.
15 Forest Knolls Estates - Decatur, Illinois - 62521
(217) 422-1002 / 1-888-GHOSTLY
Visit us on the internet at http://www.darkhavenentertainment.com

First Edition -- March 2008
ISBN: 1-892523-57-4

Printed in the United States of America

THE HAUNTED FIELD GUIDE SERIES

Welcome to the new book in a continuing series from Whitechapel Press and Dark Haven Entertainment that provides our readers with field guides to not only haunted places, but to the how-tos, mysteries and riddles of ghost research, as well. In this book, and the books to come, we will continue to take you beyond the edge of the unknown in pursuit of Americas haunted places and continue our efforts to explore the many aspects of paranormal research.

We hope that you enjoy the series and that you will journey with us in the future as we take you past the limits of American hauntings and beyond the further reaches of your imagination.

Happy Hauntings!

TABLE OF CONTENTS

Foreword by Rosemary Ellen Guiley -- 7

1. Paranormal Technology -- 9
2. The Basics -- 12
3. Looking for a Grain of Evidence -- 16
4. Digital Dimension -- 23
5. Battle of the Brains -- 29
6. Invisible Light Photography -- 36
7. Contaminating Evidence -- 43
8. Video -- 53
9. Matrixing -- 61
10. ITC -- 64
11. Sound Recording -- 66
12. Electronic Voice Phenomenon -- 74
13. Direct Radio Voice -- 83
14. Psycho on -- 88
15. Spiricom -- 91
16. Ghost Boxes -- 95
17. Ghost Detection -- 103
18. Other popular Devices -- 109
19. Buildable Paranormal Gadgets -- 114
20. Working with a Sensitive -- 125
21. Into the Future -- 128

Glossary -- 130

FOREWORD

By Rosemary Ellen Guiley

Ghost hunting has never been better – or more interesting, productive or fun. The reason is tech. Today's paranormal investigator has available an impressive array of highly sophisticated equipment, much of it at reasonable cost. Technology that puts people into space, powers the military and the Internet, and oils the workings of society can now be used in the search for evidence of ghosts.

Not long ago, the average ghost hunter sat in the dark and hoped for the best. Paranormal investigation was largely a subjective, eyewitness experience, augmented occasionally by photographs and audio recordings. Less than 100 years ago, British psychical researcher Harry Price revolutionized the field with his version of high tech: felt overshoes (for quietly padding around), steel tape measures, string, electric bells (for motion detection), a film camera, a remote-control movie camera, mercury (for detection of vibrations), fingerprinting equipment, telescope, portable telephone, chalk, and other items. Today, Price would be amazed at the outfitting of the average investigator.

The old eyewitness techniques still hold their value. No one disputes the excitement of a firsthand paranormal experience. But having a piece of hard data to back it up is even more exciting.

Tech has turned a largely amateur pastime into a professional field. The tools of the paranormal trade have helped to fuel reality shows, documentaries and films, and conferences. For the past decade or so, paranormal investigation has exploded in popularity.

The choices investigators have for equipment are many, and often not easy to make. Having a lot of great gear does not necessarily make you a great investigator. How do you know which model of a camera is best? How should an EMF meter be used? How do you protect the integrity of your data? And how do you evaluate what you captured?

You're holding one of the best books to guide you in all those considerations. Over the years, I have been privileged to work with some of the leading tech experts in paranormal investigation. Craig Telesha is one of the best of the best, and one of the first persons I turn to when I need advice, consultation and evaluation. Years and years of firsthand experience, experimentation, innovation and invention have given him a great depth of knowledge. Craig doesn't just talk ghost tech. He *is* ghost tech. He knows equipment inside and out: how it's supposed to work, how it really works, and all the pros and cons. He is constantly testing, testing, testing.

In this book, Craig guides you through tech explanations and some fascinating history. He evaluates a wide range of techniques and tools, and offers tips on uses and on data analysis. Whether you have a few pieces of equipment or many, this book will enable you to get the most out of what you use, and the most bang for your budget buck.

You don't have to be a tech expert yourself to understand it all; Craig explains everything in clear and simple terms. Beginners as well as seasoned investigators will find this an essential companion. Get it. Read it. Use it. It will be one of the most dog-eared volumes in your paranormal collection, a sign of a truly valuable book.

In the ghost hunting explosion, there seems to be a rush to produce evidence. For example, websites are full of dubious photos and electronic voice phenomena clips. Granted, a lot of evidence is "inconclusive" and subjective interpretations are involved. Even so, the best investigators, like Craig, have a critical eye and ear. This book will help you improve your analysis skills.

People have been undertaking investigations of paranormal phenomena for nearly two centuries now, and still we have no conclusive proof of the existence of ghosts or survival after death, at least in a scientific sense. Nonetheless, investigators have produced an impressive mountain of evidence that supports the truth of both ghosts and survival. As our technology improves, we may have increasingly better odds of capturing the elusive hard proof. These are exciting times for paranormal investigators!

Perhaps you're drawn to ghost hunting out of personal interest, or perhaps you want to make serious contributions to the field. Either way, this brilliant book will serve you well.

– **Rosemary Ellen Guiley**
Paranormal investigator and
Author, ***The Encyclopedia of Ghosts and Spirits***

1. PARANORMAL TECHNOLOGY

Technology is an ever-expanding and changing industry which seems to leave many people with the illusion they are getting old or they're just not savvy enough to comprehend. Everything is getting smaller and more complex on what appears to be a minute-by-minute basis. The paranormal field has grown quite rapidly over the last 15 years with the introduction of digital cameras, recorders and video equipment that boast many different options and settings that could send even the most sophisticated user into a cursing fit while attempting to operate the device during an investigation. Many user manuals that accompany these devices give pretty good instruction on how to activate the features, but most do not explain what the features really do and obviously do not cover how they may be used to capture evidence during a paranormal investigation.

Fortunately, you are not really a techno ninny, you're just missing out on some education. Ever get mad that a five-year-old can pick up an electronic device and immediately start using it, and you are lucky if you can figure out how to turn it on? The thing that separates some of the techno nerds from the not-so-technical is what I like to call the break factor. I myself have been into technology since I was five years old, but back in those days, it wasn't really about knowing how to use a device, I was just more likely to push buttons until the product either did something or wound up breaking. Having the want, need and desire to learn how something works is what drove me to tearing everything I owned apart to see what it looked like inside and trying to figure it out. I don't think my Furby made it more than three hours before it was skinned and in pieces on my desk. Obviously you do not need to go to these lengths to learn about a product, but what's the harm in pushing a few buttons to see what happens? I think most people are scared that if they hit the wrong button, the product is going to explode into a million pieces.

Pushing buttons is typically not a bad thing and you are more likely to learn about a product than you are to destroy it. I was a preteen when my family bought our first VCR. The joy of recording video and playing it

Furby having a bad day

back later was a novelty and I did it quite a bit. One feature of the VCR allows you to fast forward through the commercials while still seeing a picture on the TV, although the images speed by at an alarming rate. My mom would give me tons of grief when I would do this, saying I was going to break the VCR because, "It's not designed to do that." My rebuttal was always, "Well it would not do it if it wasn't designed to." In reality, it *was* designed to do that; it just looked harmful since the picture was barely watchable and there was no sound, so she just assumed it was bad. Though the manufacturer designed the feature for exactly what I was using it for, it wasn't very clear in the manual, so neither I nor my mom had a leg to stand on when it came to debating the scan feature of the VCR. Do you really think a manufacturer would put a feature on something that is going to break the product? In today's world of microprocessors, it is almost impossible to break a device by pushing the wrong button at the wrong time, in most cases the device will warn you that you can not do x when y is happening.

Unfortunately it is impossible to cover every single device ever made, but hopefully throughout this book I will be able to give you the logic and knowledge to be able to understand the instructions that came with the device in question and also instill a bit of comfort that they are not as complex as they appear. As you learn the standard features of your device, you will actually gain a good working knowledge of standard features of similar devices. Just as a stop sign in Maryland looks the same as a stop sign in California and means the same action is required before continuing, most comparable devices share the same terminology and symbols, which even cross over to other devices that are much different than your own. There really is a rhyme and reason to this madness. You do not need a Ph.D. or an understanding of what is going on inside the devices electrically to be able to use them like a pro. It really isn't rocket science, though some of the technology may be derived from it. We will also cover the paranormal aspect of these devices and, hopefully, shed some light on how they can be used during investigations.

2. THE BASICS

Paranormal investigation is a wide-open field that encompasses everything from ghosts to UFO research and beyond. In a technical and scientific sense, paranormal is defined as something which defies science or is beyond normal scientific explanation. The normal scientific community is typically against studying the paranormal based on several reasons, such as there is nothing tangible to study, to put under a microscope or be tested to get hard factual results. Though this may be true, there are portions of normal science we can use to help substantiate the theories of paranormal science. Don't let the word "science" scare you, we are not going to be involved in complicated mathematic equations or building evil torture devices in the basement, but we do want to refresh our memories of the electromagnetic spectrum.

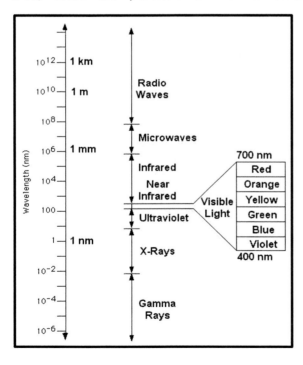

You may remember this discussion from grade school and you may have studied it further in high school, but after that, you probably had no use for the electromagnetic spectrum in your

Strange Frequencies - Page 12

normal everyday life and most likely have forgotten about it. This is a portion of science that affects all aspects of our lives, from what we see to sending e-mails to talking on the phone. We want to take a close look at this spectrum, as I will be referring to it throughout this book. It will help in explaining such devices as cameras, digital recorders and the ever-popular EMF detector. Since people learn different ways, if you find my explanation is not cutting it for you or you just want to learn more about the electromagnet spectrum, there is plenty of information available on the Internet by searching for "electromagnetic spectrum."

If you were to toss a stone into a still pond, the result would be a wet stone that sinks to the bottom, but on the surface, waves will radiate outward from the point where the stone entered the water. Energy exists at a wide range of wavelengths which resemble the waves on the pond. You can easily change the speed of the pond waves by throwing in another stone from a different height, which will change the size of the waves and how rapidly those waves move from the place the stone entered the water.

The electromagnetic spectrum has various units of measure, but we are going to deal with it as meters since most people are familiar with this unit. Just like the waves on the pond, all electromagnetic waves have a peak, a valley or a crest and a trough. The length of one wave is measured from crest to crest, so if you could measure one pond wave from crest to crest in centimeters, you could figure out its length or

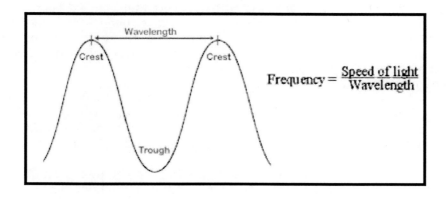

$$\text{Frequency} = \frac{\text{Speed of light}}{\text{Wavelength}}$$

wavelength. Knowing the length of the wave and the speed of light, we can then do a simple math problem and get the frequency of the wave. Over a short period of time, the pond waves will taper off to nothing, but electromagnetic waves stay constant at the same length and frequency.

Knowing this information, let's look at an electromagnetic wave you are dealing with right now as you read this book. Light is a wavelength that is also measured in meters, but this unit is much smaller than what you will find on the average ruler and is measured in the next shorter unit from millimeter, which is nanometers (nm). The longest wave the human eye can see is red, which is about 700 nm and the shortest wave is blue, which is about 400 nm, with all of the other visible colors somewhere in between. We see color based on how light bounces off an object; some of the light waves are absorbed where others are reflected. The waves reflected from the object are what we see and what we perceive to be the color of the object. White bounces the most light off an object and black bounces the least.

Most of the waves we will deal with are longer than what our eyes are able to see. The next longest wave on the spectrum is "near infrared" light, which some people are able to see just a little of. It may be viewed as a deep red color, but for the most part it's invisible. Near infrared covers from about 700 nm to about 1500 nm, give or take. This is up for debate as some consider it a broader range and some smaller. Near infrared is the same light Sony Night Shot camcorders and Infrared 35mm film are able to capture. Below near infrared is infrared, which is more heat than it is light. infrared is from about 1500 nm to 1,000,000 nm or 1 mm. You have most likely heard of thermal cameras which can see infrared, but without this type of technology, you can at least feel it as heat.

Below infrared the waves can no longer be viewed as light but can be picked up by various devices like EMF detectors. In most cases, you will hear about these waves in frequency instead of wavelength so the short math problem listed earlier is used to convert their values. Microwaves, which are directly related to your microwave oven, are just below infrared. These waves are about 1 mm long to about 30 cm long. Microwaves produced by your oven are on the longer end of these waves and are about 29 cm long. The shorter microwaves are used for things like military and commercial radar.

The longest waves listed on the electromagnetic spectrum are radio waves such as AM and FM and television, these waves can be as short as 30 cm and as long as a football field. It's weird to think about all these waves bouncing around us on a constant basis and we cannot see any of them except visible light.

So what's above what our eyes are able to see? The next step above is ultraviolet light, which is from 400 nm to about 10 billionths of a meter.

Above UV you will find X-rays (10 billionths of a meter to about 10 trillionths of a meter) and then gamma rays, which have wavelengths of less than about ten trillionths of a meter. We will hit on ultraviolet just a touch, but for the most part, the waves we will deal with are below the human eyesight range.

By now you may be wondering where you fall on this scale of the electromagnetic spectrum or if you fall onto the scale at all. It is known that the body produces an electromagnetic field, but unfortunately our bodies are not constant and our brains do not produce a constant wavelength. However, it is known that the wavelength is quite long and falls down below the AM radio band. This obviously varies from person to person and can vary quite greatly in a single person at any given time.

Though this science has been around for hundreds of years, it has been expaned from time to time and it should be safe to say we do not know everything there is to know about electromagnetism. From a paranormal perspective, the big question is, where do spirits fall on this spectrum?

3. LOOKING FOR A GRAIN OF EVIDENCE

Cameras have been around for over a thousand years, long before the creation of chemical based films and before permanent snapshots had even been thought of. In ancient times, a large enclosure with a small pinhole to the outside world would project an image on the opposite side of the enclosure that could be used to recreate the external image by tracing its outline and filling in the detail with paintbrushes and other instruments. This enclosure is known as a "camera obscura" and is the foundation for photography as we know it today.

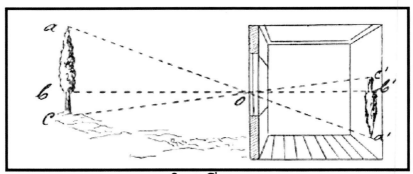

Camera Obscura

It's hard to believe photography is so simple, but anyone with a shoebox, a thumbtack and a little time on his or her hands can create a mock camera obscura to see how it works. Simply punch a hole in the middle of one of the long ends of the shoe box and head outside on a sunny day. Flip the box upside down and look at the inside opposite from where the hole is and you can start focusing in on nearby objects. This would be the de facto way of capturing images for centuries until the creation of photosensitive compounds in the early 1700s.

It all started when Professor J. Schulze mixed chalk, nitric acid and silver in a flask during an experiment and discovered that the side exposed to sunlight turned darker than the original compound. But just like any great discovery, testing and changing ingredients would take about another hundred years before images could be permanently captured.

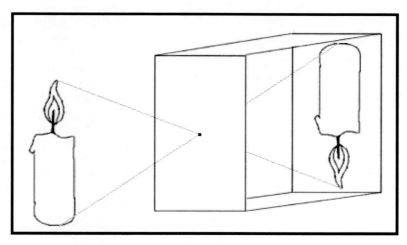

Shoebox Camera

The first photographic images were silhouettes of objects placed on photosensitive paper, which was black and white and looked like what we think of as a negative today. In 1816, Nicéphore Niépce figured out how to combine a camera obscura with photosensitive paper to produce the first photographs, but they were still negatives. It wasn't until 1834 that Henry Fox Talbot figured out how to create a positive from a negative. His technique allowed for many positives to be created from one negative and the technique known as photography was born.

The word "photography" was coined in 1839 by Sir John Herschel and derived from the Greek words for light and writing, which obviously still describes the process as we know it today. Although the chemicals have advanced since then, we are still using the same techniques for our own portraits. Film companies are always working on better solutions to a process which started over 200 years ago.

Film cameras are a dime a dozen these days and new models emerge on the market almost daily boasting a plethora of ease-of-use options for the typical shutter bug and various lens attachments and manual settings for the diehard film photographer. Unfortunately, Niépce and others who were working on this technology in the 1800s did not have auto zoom and auto focus options or even rolled film. In fact, they were

lucky simply to have something as crude as a shoebox camera obscura with a lens attached to it that did not even include a shutter.

The first commercial film was on large plates that only held one picture. These plates were inserted into the back of the camera with a second plate covering the exposure area until the scene was ready. The camera itself was a large box with a lens on one end and a place for the film to slide into the back. To take a picture, the secondary plate covering the film was removed to start the exposure and replaced once the picture was complete. Shutters were not added until the late 1800s, when film exposure time reached a point where it only took seconds to take a picture.

Using the shoe box camera obscura as our model, we can see the external image is projected upside-down onto the back side of the box. If we were to slide a piece of photo sensitive paper right where the image is being projected, we could, in fact, take a picture with our shoebox. During the exposure process, the light and dark areas of the image actually burn themselves onto the paper by changing the color of each individual grain of the photosensitive paper based on the amount of light each grain is getting. Lighter areas will turn the grains darker than dimmer areas. So basically, blacks barely burn, leaving the grain unscathed or producing white and white really burns the grains, turning them to black. This is why we have a negative before we have a positive.

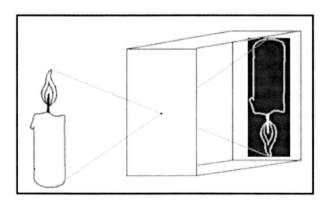

To create a positive, we actually do a similar process, but instead of using the original image projected through the shoebox, we use the negative we just created, which is laid on top of another piece of photosensitive paper and exposed to light through the negative. Again, the whites will turn black and the blacks will turn white and

Negative to Positive

since we are using a negative that is a reverse of the original, we will now have a positive.

What makes the film so important is the very process we have been discussing so far in this chapter. Back in the day when Sir John Herschel and other pioneers were trying to perfect the photographic process, many different materials and chemical compounds were used to try and produce an acceptable final product. Various bonded metals, paper and even leather made up some of the experimental materials but the compound coating used for the actual photo process is what makes up the quality of the final product.

The process was quite slow and cumbersome and even the initial snapshot could take up to 20 minutes of sitting in front of a camera trying not to move as the film was being exposed. This is why pictures from the 1800s depict subjects who appear to be either depressed or maybe even angry. Since the process was so slow, it was much easier to sit still with one's mouth closed than to smile for 20 minutes straight. Today, we use plastic coated with photosensitive material, which consists of microscopic light-sensitive silver-halide granules. There are different levels of granules that make up how sensitive they are to incoming light and how fast the image is set into the granules. This is better known as the "speed" of the film.

Film speed is rated on a scale called ISO. Though there is a ton of mathematics involved with determining the ISO, for the general consumer we really only need to worry about the final number. The ISO number is how fast the film reacts to light that is coming through the lens during a picture. The lower the number, the slower the film is to react. The slowest film is 25 ISO and the fastest to date is 3200 ISO. Each has a myriad of uses depending on the situation. Professional photographers pick their film based on their subject, so if they are shooting immobile objects under sunlight, they may choose a 25 ISO film whereas if they are shooting still subjects in dim lighting, they may choose a 200 or 400 ISO film. The ISO also depicts how quickly the image is going to get imprinted, so if a photographer is trying to take pictures of cars racing at 100 miles an hour, a 1600 ISO film will take a nice clear picture whereas a 25 ISO would show up as one big colored blur.

Though it may look as if using 1600 ISO film is better due to its quick speed, there is a trade off. In general, the slower the film, the better the quality. The difference is the size of the silver-halide granules used to produce the film. With 25 ISO film, the granules are very small and the final product will look extremely polished and clean, but a lot of light is required to get a good picture and obviously any quick movement while the picture is being taken may cause a blur. The higher the ISO number, the larger the granules need to be to make the reaction time faster. Larger granules lead to spottier grainy pictures and slightly lower quality.

Cameras that use 35mm film are the de facto standard these days. Most produce an excellent quality picture and are easy to use. There are two basic types of 35mm cameras available on the consumer market: point and shoot and SLR. If you talk to any professional photographer, point and shoot cameras are just toys and SLRs are the real deal. Unfortunately, most of us cannot afford even a cheap 35mm SLR camera,

25 ISO (left) 1600 ISO (right)

but what really is the difference?

SLR stands for "single lens reflex." In a nutshell, SLR means you are able to see through the camera lens when looking in the viewfinder to line up your shot. You see basically what the film will record when you push the shutter button.

With a point and shoot camera, you look through a separate viewfinder, which is normally just a hole through the case. Your shot will not be as accurate as with a SLR and you may not see a glare, sun streak or finger in front of the lens.

SLR is what allows the camera to have interchangeable lenses and manual focusing. Without the SLR, you would not be able to see what you are focusing on through the lens. Most SLR cameras give you full control over the settings, but this feature is becoming more common in point and shoot cameras as well.

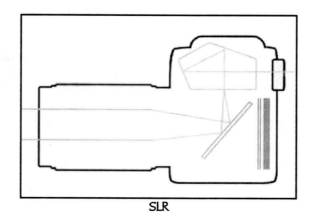
SLR

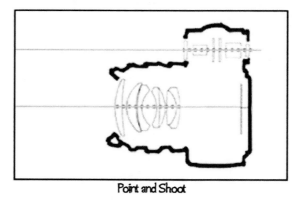
Point and Shoot

When dealing with the paranormal, SLRs kind of lose their luster as we are not trying to perfect the framing and angle of a shot. In most cases, we do not have the time or even the lighting to see exactly what we are doing. This is where things get difficult to figure out which camera is really the best for our situation. Selecting a camera is not always easy, but since there is no camera created specifically for ghost hunting, you really have to take a logical approach as to how you make your decision.

Obviously, SLR is going to take a higher-quality picture and it will definitely have more features than a point and shoot, but most of the features are not very useful for nighttime photography. If you are lucky enough to be able to afford an SLR camera then don't hold back. Make sure it has programmable settings for the focus, F-stop and iris that will allow you to find the right setting and lock them into the camera for ghost hunting. I would suggest getting one with a hot shoe on the top for adding an additional flash, which can make a huge difference and also help cut down on external contamination.

If you are unable to afford an SLR camera, then a point and shoot will work just fine, but you must be careful with your selection as they are not all created equal. When it comes to the paranormal, the actual size and construction of the camera can make a huge difference. Many point and shoot cameras have programmable settings just like an SLR. Be sure the camera you choose has this ability. Try to stay away from mini compact cameras with a flash right next to the lens. As covered in the contamination chapter coming up, the closer the flash is to the lens, the more likely you are to get contamination in your shots. It is best to find one with a pop-up flash and if at all possible, a footprint that resembles an SLR.

4. DIGITAL DIMENSION

Digital cameras have been taking the market by storm and in some cases pushing the film camera out of production. When digital cameras hit the paranormal community in the mid '90s, many groups were quick to dismiss their usefulness for investigations and some even claimed digital cameras cannot take pictures of spirits. This is one of the very statements that got me interested in ghost investigations. I wanted to know why and how someone could make such a claim. I was unable to get a suitable answer that made any technical sense to myself or based on any kind of factual information. I quickly found many folks were simply misinformed about this new technology, which sent me on my own investigation.

For starters, most people believe digital cameras were created in the late eighties and started hitting the consumer market in the early nineties, but do some digging on the subject and you will find the very first digital was actually created in 1975 by Kodak Engineer Steve J. Sasson. The unit was very large and weighed about eight pounds. The pixel resolution was only 100x100, which in terms of today's digital cameras is only .01 megapixels.

Unfortunately, back in the seventies, the only media available for saving data was large cassette tape-type drives. The camera took a whopping 23 seconds to write an image onto the tape. It was very impractical for the time, but at least it only took 20 years before it was available to the public, unlike film, which took over 200 years.

First Digital Camera

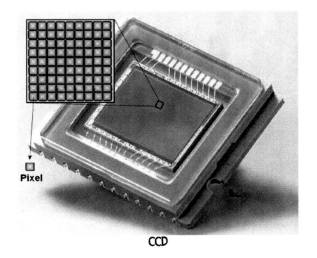
Pixel
CCD

The first consumer grade digital cameras that emerged onto the market in 1990 were black and white and boasted a measly .032 megapixels with one megabyte of memory, which could only store 32 pictures. The cost was a mere $995.00. Believe it or not, the professional market already had a 1.3 megapixel SLR camera with 200 megabytes of memory, which cost $30,000 for the basic kit. Each year the megapixel count inched up just a bit with new models being released every so often. Just like film, it was a slow start out of the gate for digital cameras until the technology and quality were at least fair enough for the general consumer and obviously, until the price dropped enough to be affordable.

Digital cameras are definitely not as simple as film cameras when it comes to their technical explanation. Instead of film they use an electronic device called a CCD, or charged coupled device.

For the most part, it's a bunch of tiny little light sensors, each smaller than the head of a pin, arranged on a square grid. You may be more familiar with this technology than you realize, but finding similarities in everyday products is not always evident. You either have or know someone who has a nightlight in the bathroom that turns itself on when it's dark and off when you turn on the overhead light. This device uses a light sensor. Although it's only one and hundreds of times larger than a single sensor in a CCD, it is basically the same thing. If you play with the light in the bathroom by slowly covering up the hole where the light sensor is, the light will slowly come on and slowly go off when you remove your hand. What you are essentially doing when covering the hole is changing the electrical flow through the sensor, which changes the electrical flow to the light bulb.

To take this explanation a bit further, let's say we bought 625 nightlights and arranged them in a grid of 25x25. If you were to put your hand in the middle of the grid, the lights under your hand would illuminate and create an image resembling your hand. Now if we combine this idea with the shoebox camera obscura and if we could shrink our 625 nightlight grid to fit in the shoebox at the back where the image is projected, we could make the lights create an image of whatever is projected into the box. This is a very crude example, but this is how the CCD in a digital camera works. Each tiny light sensor is called a pixel and is how the cameras are rated in clarity. "Mega" stands for one million, so one megapixel is one million pixels.

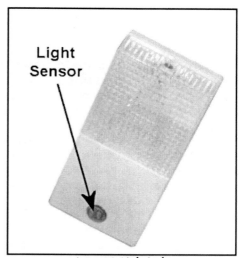

Automatic Night Light

At the time of the writing of this book, the largest megapixel camera available to the consumer market is 12.8 megapixels. If we wanted to reproduce that with nightlights, we would need 16,384,000,000 (that's 128,000 x 128,000 pixels on a square grid). If you are wondering how that compares to the amount of granules on a piece of 35mm film or what the megapixel equivalent

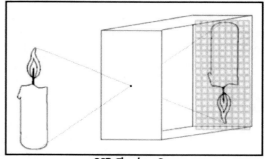

CCD Shoebox Cam

of an SLR 35mm camera is, well, they are pretty comparable. A good quality film will be equivalent to about 12 megapixels, so digital cameras have reached the point where they are comparable in quality to a 35mm camera. Obviously, there are many variables to this, but as a whole, they are pretty even at this time.

Now that you have an idea of how a digital camera sees the image it's recording, let's look at the rest of the process. When you push the shutter button on a digital camera, the light sensors are energized and the camera reads the value of each light sensor and saves that information to the camera's memory. Just like a film camera, the initial picture is read in as a negative, but since we are dealing with a minicomputer in our hands, the information is changed to positive before it is written to the memory. There is plenty more going on while the camera is writing to the memory, such as the image being flipped and mirrored so it represents the actual scene. But for our purposes, we are only concerned with everything that happens before and just after the CCD was read by the camera and, of course, the final product.

Digital cameras conform to the same ISO standard as discussed in the film camera chapter. The ISO settings dictate how sensitive we are asking the camera to be to incoming light. Unlike the film camera, the digital has to compensate for these settings in a much different way. During a higher ISO picture, the incoming light is amplified to the equivalent of the setting. The standard everyday picture setting is ISO 100, just like with a film camera. If you set the camera to 200 ISO, the incoming light is amplified by two. If the ISO is set to 400, then the light is amplified by four, and so on. This amplification is not much different than turning up the volume on a car radio to make the sound louder. But once you get the volume up to a certain point, the car stereo sound will start to crack and distort due to being over-amplified. A similar problem happens with a digital camera when using a higher ISO setting: amplification creates speckles and grain in the final picture. So the same holds true for digital cameras as in film. The higher the ISO, the faster it's going to record light but the grainier the picture will be.

Finding the right digital camera for your needs is a headache waiting to happen. There are so many different models, megapixels, styles and features it will make your brain spin. Just like film, there are two different types: point and shoot and SLR and both have a myriad of features that most consumers will never use.

Before we get into the different types of cameras I should also mention another type of sensor which works much like the CCD and is quite common on today's market. When searching for a camera, you will come across models with a CMOS sensor, which is not much different than a CCD but it is slightly different in its internal workings. Back in the day, CMOS sensors were considered lower quality than CCDs, which was typically used as a selling point. As of 2001 or so, this was just not true anymore and CMOS sensors have become just as good if not better than their brother the CCD. Unfortunately, there are still many salespeople out there with the antiquated belief that CMOS is inferior to the CCD, so if they are using this as a selling point to push you into another camera, let them know that you know better.

Another point I need to mention is the amount of megapixels a camera is claiming to have as to how many it's actually using. You may see cameras that claim to be 8 megapixels but in the fine print says 7.2 effective megapixels. What this means is some of the pixels, which are normally around the outer edges, are either unused, not visible to the lens or are used for color correction and other functions. So the resultant picture is actually 7.2 megapixels. This is not a huge issue but definitely worth noting.

If you can afford it, a SLR is definitely the way to go, but you are looking at a pretty salty price just for the camera. Some models require you buy a lens separately as well as the memory on which to save the pictures. A $700 price tag could quickly turn into $1,500. You may be thinking your old lenses from your 35mm SLR might work on the new digitals, but this is mostly untrue though some do have adapters. The camera should have the ability to be set up manually and if you are spending this kind of money, get one with a hot shoe on the top for an external flash. This will prove to be very useful with paranormal investigations as well as normal everyday snapshots. Some of these cameras boast a kind of Night Shot or IR photography. If you can handle it, get one with this feature, although I think Sony is the only manufacturer currently making this kind of still camera.

For the rest of us, point and shoot will work just fine, but these can be trickier to select than an SLR due to the amount of choices available. Smaller size is always appealing and seems to be what the manufacturers are working towards. When ghost hunting, this will prove to be more of a pain with flash pictures than you may want to deal with. The flash location is very important. Try to look for something with a pop-up flash that sticks up from the top of the camera as opposed to a built-in flash that is situated just above the lens. The closer the flash is to the lens, the more contamination you will get when taking flash pictures. We will cover why the flash location is important in the contamination chapter and what you can do to help rid yourself of contamination in photos.

For both cameras, you will want to be able to set up the focus, aperture and F-stop settings manually. Most cameras have many other features, including low light settings and some even come with infrared capabilities. These two features would be a nice addition if you can handle the cost. Removable memory is also a plus and allows you to take along extra memory cards in case you run out of room on the initial card. At last, you get down to the final decision, which is how many megapixels do you need?

Selecting the megapixels is a tough choice and is really going to depend on what you plan on doing with the pictures. Obviously, the more pixels the better the picture is going to look, but there is a point where you may be just wasting space. Fortunately, most cameras have a feature that allows you to adjust the resolution, so a 12.8 megapixel camera can take pictures at only 5 megapixels. Personally, I would not get anything lower than 5 megapixels and would most likely go with a 7 or 8 megapixel camera, but if you are willing to put up some cash, it's not going to hurt to go higher.

Be cautious of no-name cameras and ones combined with other devices. Most cell phones come with digital cameras built in, but the quality is average at best. Some video cameras also boast still pictures as well, but again, the quality is normally nowhere near that of a standard camera and you do not want to have to stop recording video to take pictures.

5. BATTLE OF THE BRAINS

Preparing your camera for a paranormal excursion can be cumbersome and time-consuming. Most newer film cameras and all digitals have an internal computer brain with many settings that have been programmed for different lighting situations. This is typically called auto mode and for most standard snap shot situations, they do a pretty good job and the pictures turn out great. However, we are not in a standard situation and even though there may be several preprogrammed settings on the camera to choose from, none of them say "ghost," "paranormal" or "pitch black." The main culprit we have to deal with during an investigation is light or lack thereof. In most cases, we will be ghost hunting at night with all the lights out. Just as we are basically blind in complete darkness, our cameras are just as blind, which makes it difficult for the computer brain to make a good logical decision as to what settings it should select to get the best picture. Due to this dilemma, it is best to set our cameras up manually. This will take a little time but is well worth the effort.

To get yourself familiar with what you are attempting to accomplish, you will want to experiment with the camera. You can do this in your own house with all the lights off to simulate being in a haunted location at night. Be sure to write down each setting as you take pictures so you will know which ones look the best when you view the final picture. I would suggest a 400 ISO or 800 ISO setting for this; 1600 will most likely be overkill. Obviously, if you are setting up a film camera, you will need to purchase a couple of rolls of film with the correct rating. You don't want to get too much grain in the picture or you will wind up producing matrixing effects, which we will cover later. If you are using a digital, adjust the ISO accordingly and reference the manual for how to get into the settings and where to make the change.

The following information will work for both SLR and many newer point and shoot cameras, both digital and film. You will need your instruction manual so you know which setting we are talking about and also how to program the settings into the camera. Don't let the word "program" scare you. This does not mean you need a Ph.D. and it's actually much simpler than you may expect. In doing this, you will learn several things about your camera as well as what pictures will look like at different settings. Take your time and document everything. Once you are done, you will have a nice reference chart to go by. I would suggest using a tripod so the scene is consistent. Set the tripod in an area which will frame as much of the room as possible.

First let's cover the focus, since this setting should remain constant for investigation purposes. Most cameras have a setting called "infinity" which is where we want to set the focus from here on out. What this basically means is everything will be in focus far and near, which will give us the most extreme wide angle shot the lens is capable of. The infinity setting will most likely look like a sideways 8 which is the scientific symbol for infinity. Although most cameras have auto focus, it is very unreliable in extremely low light   . A static focus of infinity is the best solution. If the auto focus continues to work after you change the camera to infinity, look for a setting that will allow you to put the focus into manual mode.

Focus set to Infinity

Next, we want to look at the shutter speed. This one can be kind of tricky and not all settings are on all cameras so you will want to reference the manual. The shutter speed will be a bunch of numbers from possibly 1/2000 to 15 and maybe even higher or lower. The numbers are actually the fractions of a second until you get to 1, then they are full seconds thereafter. So, to keep it straight, the 1 is actually 1 second, 1/2 is actually one-half of a second, 1/4 is actually 1/4th of a second and so on. An average sunny day picture should have a shutter speed of about 125th of a second (1/125 with 100 ISO film.) Since we are using 400 ISO let's start there. We must be careful not to drop the shutter speed too low or the pictures will start smearing.

The next setting we will want to work with is the aperture. This is also known as the F-stop and it functions just like the iris in your eyes. Its basic purpose is to keep light at a constant level regardless of how dim or bright it

Shutter Speed

is. When there is a lot of light, the iris will close to allow only a little in. When there is very dim lighting, it will open up to let more light in. The F-stop is based on a number system just like the shutter speed and typically goes from 1.4 to 22. The smaller number of 1.4 means the aperture is open all the way and letting in as much light as possible, 22 is as small as it can go and letting in as little light as possible. Some newer point and shoot cameras have a broad range, but most have only a couple of F-stop settings. Let's start at an F-stop of 1.4 (all the way open) or the lowest number available on your camera.

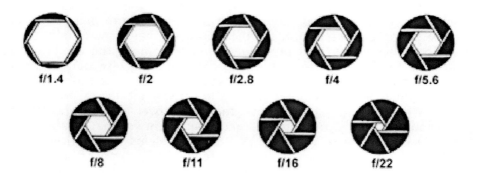

The last setting to activate is the flash. This is normally indicated by an icon of a lightning bolt. If your camera has a red-eye reduction feature, turn it off. No sense in wasting batteries on something we are not using. If the camera has a date and time feature that allows you to stamp it on the picture (for a film camera and some digitals), turn this feature on so we can use it to identify which picture is which. So now the camera should have the following settings: the focus should be set to infinity, the shutter speed should be set to 1/125, the F-stop (aperture) should be set to f/1.4 or the smallest number on your camera and the date and time feature should be on, as well as the flash. Write down your settings along with the number of the picture and the time. You are now ready to take the first picture!

Take the picture, then increase the aperture to the next setting (next higher number) and write down the setting. Take the next picture and increase the aperture again, as well as writing down the setting. What you are doing is slowly decreasing the amount of available light to find the right setting. Once you run out of F-stop settings, move the F-stop all the way back to the starting point (1.4 or the smallest number available on your camera), then decrease the shutter speed by one setting. This will most likely be 1/100, which is one 100th of a second. Repeat the procedure going through the F-stops again. Don't go too much lower than 1/30th on the shutter speed as it will get to the point where the shutter is too slow and handheld pictures will smear or blur.

Once you have taken all the pictures, get your film developed or download the pictures to your computer and see which ones look the best then program the camera for those settings. Once you have everything programmed you will be ready to take good-looking pictures during an investigation and not have to worry about the focus going crazy or the camera's brain making a bad decision based on a reflective surface in the room, etc.

Most cameras will hold these settings in the manual mode almost indefinitely until they are altered by human intervention. Just be sure each time you fire up your camera that the settings are what you expect before taking pictures. Another precaution you may want to take before heading out on your first investigation is to remove the camera strap and lens cover, complete with string and anything else dangling from the camera. If you are worried about scratching the lens and don't want to leave the lens cover behind, remove the string and take the lens cover itself. While taking pictures, put the lens cover in your pocket for safekeeping. Take extra batteries and make sure you have backups for your backups. One of the most common problems experienced in the field is quick battery drainage. Make sure you have plenty of batteries!

Taking pictures in the field is not a terribly difficult process but you will quickly find the normal photography rules don't really apply when sitting in a pitch black room with nothing to frame in your viewfinder. This makes it difficult to make a decision of what to take a picture of and when. Obviously, if you are using a digital camera, you are not going to be as worried about the amount of snapshots you will take as someone with a film camera. If you are ghost hunting with a sensitive, you may want to let them guide you as to when and where to take a picture, but if that's not the case, then you have to go with your gut feeling.

Other tools discussed in this book may be able to help, such as an EMF detector or temperature probe. Placing these devices in the middle of the room and waiting for an indication to start taking pictures may increase your chances of getting something. Just be sure to take note of your surroundings and look for things that may reflect the light of your flash and cause a false anomaly. If you are standing in the middle of a room full of mirrors, then chances are you are going to get some strange light effects from the flash. Take several pictures of the same area so you have something to compare them to later. There are times when something may not look right and could possibly be an anomaly, but if you only have one picture, then you have nothing to compare it to.

The surroundings have a big impact on how your pictures will turn out, but you also have to be aware of what you are doing during the picture as well. If it is 40 degrees or colder, your breath may resemble smoke when you exhale. Be sure to hold your breath for about five seconds before taking the picture. Other issues can involve your clothing, such as strings hanging from hoodies or a sweater that can inadvertently release microscopic fibers into the air. Although this sounds very strange, a microscopic fiber floating in front of a camera lens during a flash picture can look like an anomaly.

There are several theories as to how and why a camera is able to actually pick up something we don't otherwise notice while on an investigation. The base theory is our brains are ignoring the spirits and dismissing their presence. Since the camera does not have the ability to discriminate against what is in front of it, the spirits may show up in a picture of an otherwise empty room. As hard as this may be to accept for some people, most of us have experienced a similar phenomenon in our everyday lives at one time or another. For instance, have you ever lost anything that you searched for frantically only to find it later in a place you searched five times earlier? Somehow our brains overlook the obvious. The scarier experience to think about is driving down the road and realizing you don't remember driving the last two miles!

Another popular theory is imprinting, which is probably the most popular and one reason some believe digital cameras cannot take pictures of spirits. In the case of this theory, the spirit is imprinting its image directly to the film and is not visible in the room during the picture. This theory may have stemmed from the superstition among some cultures that taking a picture of a person steals their soul. In my opinion if this were true, most movie stars would have expired or at least become soulless long before finishing their first film. Of course there are some movie stars who have no soul to begin with, so obviously they would be unaffected.

The latest theory is optical light range of the eye versus the film/CCD light range of a camera. From our earlier discussion, you know the eye can see 400nm to 700nm of the electromagnetic spectrum. But what about film and CCD cameras? For a simple test, take a look at the front of any TV remote control and push one of the buttons. You will see nothing or if you are lucky and it's dark enough, you may see a slight red pulsing light. Now, turn on any digital camera or camcorder with a CCD and point the remote into the lens and look through the viewfinder while pushing a button on the remote. It will look as if a bright flashlight is being shined into the lens. You have just proven a digital camera can see light that the eye cannot.

A typical CCD can see light into the near infrared range pretty easily, but most digital cameras except those like the Sony Night Shot incorporate a filter in front of the CCD to get rid of infrared light, which causes problems with capturing color information correctly. With the filter, the camera can only see from about 400nm to 750nm, or just a little into the near infrared range. Without this filter, most CCD cameras can see from about 350nm to 1200nm. The Night Shot cameras actually have this filter as well, but it is moved out of the way when the Night Shot feature is activated. In short, a CCD can see twice as much of the electromagnetic spectrum as the human eye.

Film cameras are slightly different and are controlled not by the camera but by the sensitivity of the silver-halide granules on the film. A camera with regular 100 to 800 ISO Kodak film can see 250nm to just about 700nm of the electromagnetic spectrum. Instead of seeing into the near infrared range, a film camera can see slightly into the ultraviolet range. Not all manufacturers use the same process or chemicals to make film, so this information may not hold 100 percent true for Fuji or even Kodak as they refine their processes. This information is based on up-to-date film specifications as of late 2007. Check with your film manufacturer for more information.

In a nutshell, the given information suggests film cameras are not very good at seeing into the near infrared range without special infrared film and CCDs are not able to see as far into ultraviolet range as a film camera. It is speculated spirits reflect light that may be in the ultraviolet or near infrared range. Based on the given information, this suggests both cameras are equally important during an investigation since they both can see different parts of the electromagnetic spectrum that the human eye cannot. But this can also cause a problem.

Not being able to see the anomaly while taking a picture means not knowing its root cause, whether it is paranormal or normal. Just because you did not see it does not make it paranormal. Breath being exhaled during a picture on a cold day can cause what looks like smoke or ecto mist and may not be visible to the naked eye at night until a bright flash illuminates it for a split second. They say the eyes play tricks on you at night, which is due to inadequate light information coming in. Without the proper amount of light bouncing off the smoke, your eyes will overlook it but a camera with a bright flash will not discriminate and it will capture it.

6. INVISIBLE LIGHT PHOTOGRAPHY

Not being able to physically see what the camera is picking up has brought forth the speculation that spirits reflect invisible wavelengths of light. Many investigators have turned to using filters, special film and modified cameras for their investigations, which exposes what appears to be a whole new world mostly devoid of color.

Infrared spirit photography is becoming more popular these days but is not really a new concept. In fact, it has been around since the early 1900s. In its infancy, IR photography was a hobby requiring experimental film and long shutter speeds. During World War I, resources were put into expanding infrared film to improve its exposure time and overall clarity for nighttime aerial photography. After the war, it found its place in different industries for examining plants and other subjects although infrared film photography has remained more of a hobby for the consumer market.

Digital IR picture using a modified camera

Although we call it infrared, this form of viewable light is actually near infrared and is just below the visible light spectrum from about 700nm to 1500nm. True infrared is actually viewed as heat and not light and can be found beyond 1500nm and is viewable using a thermal imaging device, which we will talk about a little later.

IR Filter

Just about any 35mm SLR camera can be used with infrared film with the addition of a filter to keep out ambient light. Using this type of film requires a bit of skill and technique, which is a book all its own. If you are interested in using 35mm IR film, there are many different articles available on the Internet and various books in print explaining what is needed and how to produce excellent results.

As mentioned earlier, the CCD in digital cameras can also see into the near infrared spectrum. Unfortunately, IR waves affect how the CCD sees color and are typically blocked by a filter inside the camera or a special coating on the lens. This makes it difficult to produce any kind of useful results under IR conditions, though some folks have attempted to use infrared pass filters and long shutter speeds.

The main issue with screwing a filter onto the front of your digital camera is that the internal filter is doing the exact opposite and blocking IR light. What you are doing is blocking almost all light from getting into the camera between the two filters.

The camera's brain is going to get confused as to what is going on and go into low light mode. Even if you put the camera into manual and adjust the settings accordingly, you will need to use a tripod and long shutter exposure to get any kind of results. The filter will most likely render the auto focus system of the camera useless as well, which can make it difficult to get a clear picture.

There are several filter options with different ratings that block ambient light down to a certain point. These start at around 630nm, which blocks most ambient light but lets a little in at the bottom part of the visible light spectrum. The next filter in the mix is 715 nm, which blocks all ambient light to just inside the IR spectrum. Filters of 760nm are probably the most popular and block all ambient light and a little more IR light than the 715nm. Most folks will find the 760nm filter adequate for their endeavors.

The higher the filter number, the more ambient and IR light are going to be blocked. Anything over 800nm is overkill for paranormal investigating. The more light blocked, the longer the shutter will need to remain open to get a useable picture. The further you go into the IR spectrum the lower the sensitivity of the CCD. Keep in mind IR filters are not designed for spirit photography; there is no tried and true filter or rating. Cameras are not created equal and will require some testing and playing around to get the right settings.

Testing the camera with a filter in place and playing around is your best defense against false positives. Start off with your camera on a tripod without the filter and take a standard picture of the scene to get the camera focused for ambient light.

After the first shot, immediately put the camera into manual focus mode to hold the last setting. Turn the flash off and put your filter on. The focus will now be a little off and some adjustment will be required. If you are lucky, the camera may show a dim picture on the LCD screen, which may give you enough information to finish focusing. If not, you will need to take several pictures and fine tune in between.

Standard shot with no filter and sun behind to the left.

Take the next shot and be sure not to move the camera or it will ruin the picture.

You may notice some of the detail appears to be missing and although you may be wearing dark clothes, they may appear much lighter or even white. IR light does not scatter into separate colors like ambient light, so color information is pretty much lost, although the picture may be slightly red or purplish. You may find it helpful to put the camera into monochrome mode if

Filtered IR picture

available.

If the picture came out completely black or very dim, you will most likely need to adjust the shutter setting manually. Most IR shots outside in bright sunlight will require a one- to three- second shutter to produce a good picture. Once you get a nice, crisp, properly focused picture, it's time to experiment with false positives and the "Mumler Effect".

In the late 1800s, a Boston engraver named William H. Mumler discovered how to produce ghostly images on film by taking advantage of the long shutter exposure and introducing an extra subject into the background during a normal photograph session. The introduced subject would only spend about half the time in the frame as the other subjects. The final product would show a ghostly image in the background that could be claimed to be the spirit of a departed loved one.

The Mumler Effect

IR filtered photography will produce the Mumler Effect if a subject moves or leaves the frame during the picture. For the first fake shot, get a human subject in front of the camera who will be prepared to move during the picture. When the shutter button is pushed they just need to get up and walk away, no need to run.

There are two possible effects you may get with this kind of picture. If you walk towards the light, you will look like a ghost with no other anomalies appearing. If you walk away from the light, you may get a swoosh or trail following you. This is due to your clothes being brighter than the background. While the shutter is open, the brain is processing light. If it sees a change which gets brighter than when the picture was first started, it will record it. If the change gets darker, it will not record it. The longer the subject stays in the frame, the more solid they will appear in the final product.

For the next shot, have the human subject stay seated while moving an arm or their whole body in an exaggerated way, then have him hold the pose.

In this case, the subject will look half solid and half ghost and may

Fake IR Ghost (Mumler Effect)

IR Ghosting Effect

even be missing some parts depending on the motion.

IR digital photography can produce some really neat results outside on a sunny day with a filter in place. No matter how you work at it, using the previous method during an investigation at night is almost next to impossible regardless of flash and camera settings. But there is a solution, actually two.

Sony offers a few higher end models with the ever popular Night Shot mode. These cameras do an excellent job at night and, coupled with an IR filter, can produce excellent results even during the day. No need to play with the focus and long shutter settings. These cameras produce instant results.

Another option is to modify a camera by removing the IR blocker and replacing it with a transparent piece of plastic or glass which is the same size and thickness. The first camera I modified back in 2000 was an old Agfa model that boasted a whopping 640x480 resolution. Though the camera was not the best quality, it did an excellent job of taking infrared pictures for the time. I replaced the IR blocker with a piece of Plexiglas that was roughly the same size and thickness, along with a Wratten 87C gel filter, which blocks all ambient and some IR light down to 760nm.

Since the Agfa conversion, I have experimented with many other types of cameras. I find regular Sony cameras to be the most difficult to work with and Olympus models probably the easiest. This modification requires a bit of mechanical ability and technical logic to work with the various connectors and mechanisms in the camera.

If you decide to attempt modifying an existing camera, you may be able to find specific information about your model on the Internet, along with modification steps and techniques. Be sure to replace the filter with a comparable transparent piece. If you just remove the filter or use a thinner or thicker replacement, the camera will not focus properly. Also be aware that colors will no longer be recorded properly.

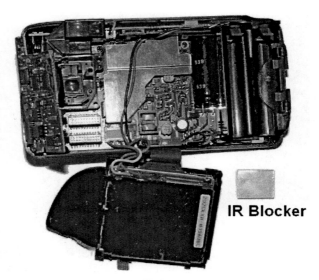

IR Blocker

Modified Camera

Working with invisible light on the opposite end of the spectrum from IR can yield some excellent photography and also may be a wavelength in which spirits are able to be photographed. In the late '70s, George W. Meek of the Metascience Foundation was working on a device to attempt spirit communication called the Spiricom. During a communication session through a medium, Meek was told "thought waves are just beyond the border of ultraviolet frequencies." In theory, if spirits retain human energy patterns after death and those patterns are just beyond ultraviolet frequencies, it may be possible to photograph them using UV light.

Ultraviolet radiation was discovered in the early 1800s by German physicist Johann Wilhelm Ritter by observing the

Agfa IR pic from Gettysburg (Sachs Bridge)

Digital UV Picture

darkening effect of silver chloride-soaked paper in sunlight. Ultraviolet radiation is just above visible light waves on the electromagnetic spectrum from 400nm to 1nm and like IR is invisible to the human eye.

Taking pictures in this spectrum is not much different than taking pictures in the IR spectrum. Typical 100 ISO film in a standard 35mm SLR camera with a UV filter can yield some interesting results. Special 35mm film can also be used to photograph in this spectrum but can be very difficult to work with. In most cases, the camera does not require a filter as the film is specifically designed to only see UV.

Digital cameras are quite sensitive to UV light right out of the box but they require a special ultraviolet pass filter for blocking ambient light and only allow UV light to reach the CCD. Finding one of these filters is another quest all its own. Tiffen 18A is the standard filter model, but unfortunately they are not very popular and can be quite expensive. Take note that UV pass filters are not to be confused with UV haze filters, which block UV light from getting to the CCD.

Working with a UV filter is just as cumbersome as working with an IR filter. This setup will require a tripod and a longer shutter setting than normal and will also mess with the focusing system of the camera. The same techniques as mentioned earlier to be used for IR can be used to set up your camera for taking pictures with UV and also show how false positives can be obtained.

Many of the techniques discussed in this chapter can also be used for video cameras and camcorders. Some cameras are more sensitive than others and some amount of experimentation will be required, but you should achieve similar results.

7. CONTAMINATING EVIDENCE

Of all of the controversial issues you will encounter during an investigation, contamination is probably the biggest and the most misleading. Contamination can produce false evidence very easily, which is partly the fault of the areas we investigate and also sits on the shoulders of the camera manufacturers. This problem plagues just about every kind of camera made today including 35mm SLRs. If it has a lens, it has the potential to be contaminated.

Not too long after digital cameras started to become affordable, the paranormal community exploded with a whole new phenomenon called the "orb." Since then, the Internet has been overrun with picture after picture of these so-called spirits hanging around sensitives and haunted locations. At first, orbs only seemed to show up on digital camera pics, but soon film cameras started picking up this phenomenon as well. Was the world suddenly getting infiltrated by spirits, or was there a more logical explanation waiting to be revealed?

I spent a bit of time and money over the last couple of years trying to come up with a solid answer as to what orbs really are and why they are appearing now and not back in 1964 when Grandpa was taking pictures of his family. At the time I started my investigation, orbs were still thought of as spirits, but some skeptics were touting the general orb picture as dust. This did not sit well with some folks and the paranormal community started making claims about which orbs were dust and which were actual spirits based on certain criteria like motion, object blockage and even color. So the controversy continued to divide paranormal groups across the country.

As I gathered cameras and information, it was amazing to find how young personal photography really is compared to the amount of years photography itself has been around. In the earlier days of personal photography, the cameras available to the consumer market were big, bulky and only had a couple of settings with most having no ability to take flash pictures. Many families could not afford them, the film speed was still kind of slow and the quality was bearable at best compared to today's standards. Onboard flash technology was in its infancy and

Early Consumer Camera

useable in semi-low light settings, but not really effective in total darkness.

Believe it or not, the early '60s brought forth the first consumer grade cameras with easy-to-use film cartridges and affordable flash technology. Kodak introduced a camera in 1963 called the Instamatic that used 126 film cartridges and created an ease of use for the general consumer, but the quality was just okay and even though flash cubes were becoming a standard, the brightness of the flash was not really too great and did not cover much distance.

In the '70s, the camera market exploded with the introduction of 110 cameras that could be purchased quite inexpensively. There was even a 110 SLR camera created by Minolta to attempt to generate sales from the 35mm SLR community. The '70s also yielded the first electronic flash type camera where the flash was built in and re-usable, ending the need for throw away flash bulbs and cubes. The overall flash range was still not great but not having to buy bulbs was definitely a plus. Also in the '70s, Konica released the very first point and shoot 35mm camera.

The '80s brought us additional refinements on many of these cameras and the famous disc camera, which was about as small as they would get on the consumer market at the time. But all of these cameras that were introduced over the years had one common trait: film. Being film-based, the amount of pictures being taken at any given time would never match the amount of pictures taken today. More care was also put into the pictures as most cameras only held a total of 24 frames and far fewer pictures were taken at night compared to today.

In the early '90s, digital cameras started emerging on the market, but the overall quality wasn't anywhere close to a 35mm camera and the case design was odd to say the least. My first digital camera experience was with an Apple Quick Take 100, which didn't even rate into the megapixels. The front of this camera was basically flat and it looked kind of like a pair of binoculars. In fact, most of the digitals of this time with the exception of the $25,000 digital SLR models had an odd look about them and did not resemble 35mm film cameras or standard point and shoots of the time.

As far as I know, the Quick Take did not have an issue with orbs. As the camera cases got smaller and the resolution got better, more and more digital cameras were taking the place of point and shoot cameras for everyday photography. The miniature size and reconfiguration of these cameras, as well as the amount of pictures being taken at any given time, is what caused the orb frenzy of the '90s

Apple Quick Take 100

As upsetting as it may be to some, orbs are nothing more than contamination of the picture by dust, pollen and other airborne particles. For the most part, these particles are microscopic and unable to be seen by the naked eye when scattered about. Make a big pile of dust and you can see it from across the room as it sits on your television or bookshelves. Though a dust rag and a can of Pledge will take care of the problem for the short term, dust is everywhere and is almost impossible to get rid of completely, even in a so-called clean room. On an average day, the ambient air outside in a typical urban environment might contain as many as 35,000,000 contaminant particles per cubic meter, a typical clean room has about 100,000 particles per cubic meter. Though it's significantly less, it still has contamination.

It's impossible to know what the particles per cubic meter would be in an abandoned haunted location, but chances are, since it has no upkeep, it's going to be significantly higher than the 35,000,000 particles outside. It does not take much more than a small unnoticeable breeze to get those particles stirred up and floating around. But this is not the only factor at play when it comes to taking pictures of dust while on an investigation. The camera manufacturers are also at fault by the very design of their devices, which started about the time digital cameras were heavily hitting the market.

The manufacturing problem has to do with the on-board flash location, which, before the digital explosion, was typically far enough away from the lens to not cause any problems. As the cameras got smaller, the flash was moved closer to the lens and started causing issues that still

Particle Contamination

plague cameras today. During a flash picture, small particles within millimeters of the camera lens will get illuminated, causing a bright round out-of-focus circle to appear on the final picture. This may be the result of dust, pollen or other microscopic particles being illuminated by the flash which would otherwise be overlooked by the camera. This problem also started plaguing smaller point and shoot 35mm film cameras not too long after digitals and it also plagues the flash-style disposable cameras that emerged on the market in the early '90s.

Looking back at our mini timeline of cameras, Grandpa would have never had a problem with contamination due to the large footprint of his bulky camera and the flash location. In most cases, he probably did not take too many pictures inside, but those he did, the light angle of the flash would never get close enough to the lens to illuminate dust particles. This pretty much holds true for most of the cameras of the early '70s and '80s. As flash technology improved, cameras got smaller and more pictures were being taken due to digital technology. The result was orb pictures starting appearing on a daily basis.

Some manufacturers have stepped up to the plate and started listing dust (orbs) as a problem in their manuals under trouble shooting. But the demand for smaller portable devices has overshadowed the design flaw plus this problem is only evident during flash pictures.

In some cases, the orbs may look like they are behind objects or appear to be emerging from walls. Some are even in motion with a kind of a swoosh behind them. Though in the past these were thought to be signs of something other than dust, it's not hard to replicate all of these phenomena in a lab-type setting. There are some simple tests you can perform to see how well your camera stands up to dust during an investigation.

These tests can be performed with a digital camera or a film camera, but keep in mind different film speeds are going to produce different results. What you will need is your camera set to automatic mode, a feather duster or dry dust rag and a tripod if you have one. Set up in a location where you can turn the lights off making the room dark. If you don't have a tripod, you can put your camera on the end of a table, but try not to move it too much so the pictures are consistent. If you have dust allergies, you may want to have someone else perform these tests for you because a good bit of dust is going to be released into the air.

The best place to get dust is on the back of your television. The high voltages produced inside attract small particles. This holds true for most televisions regardless whether they are LCD, plasma or regular tube set. To collect some dust, run your feather duster or dry cloth along the back of the TV. Do not use Pledge or another dusting agent as this will make the dust stick to the cloth and little will be released when we need it.

For the first test, leave the lights on so you can see what you're doing. If you think it will be too bright for the flash to go off, set the cameras flash to always on instead of auto. Shake the rag or feather duster a couple of times about a foot above the camera, count to five, then take a picture. If you are using a digital, you may be able to see the fabricated orbs on the LCD display screen.

Turn the lights out for the next test. Hopefully you will have enough light to see what you are doing. Again, shake the rag or duster a couple of times about a foot above the camera. Count to five and take a picture. Check your LCD if testing digital and you should see more orbs. Some of these may look like they are in motion.

The last test may require two people. You may need to reload your dust rag or duster, so run it across the back of the TV again. Hold the camera up like you are taking a picture, but this time shake the dust a foot above the camera and keep shaking away to the left or the right. Count to five and take a picture. At the same time, spin in the direction of the dust trail. Your LCD should look like it has orbs and you may even have another effect we will talk about in just a bit.

The three pictures you just took will probably look like an orb fest. Depending on where the flash is situated on your camera, some areas may look more dense than others. These are the areas getting hit with the bulk of the flash, so the orbs will look much brighter and may look solid. Areas with less light will look more sparse and the orbs will most likely be transparent, although there will be spots where they are solid.

In the first picture, the orbs will most likely look like they are stationary. You may notice you cannot see much of the background as the dust swallowed up the light from the flash. Since the camera has its own brain and we are in auto mode, we are leaving the decisions to the camera. That being said, the camera noticed it had enough light and assumed the picture was done when in fact it wound up not staying on long enough to get enough light information to show what's behind the dust storm.

In the second picture with the lights completely out, two possible scenarios may have happened. If we had enough dust, you may not even see the background as the picture would have finished before it picked anything up behind it. If you did not have enough dust, you may be able to see the room, but the orbs may look like they are in motion. The second phenomenon is actually caused by the camera not registering enough light so it left the shutter open for a little bit longer than normal. Though this may only be a 10th of a second, it is enough time for the dust to continue moving and cause what looks like a tail behind it.

The third picture should look like the dust is moving and, depending on how the camera picked it up, you may not even be able to tell it was dust. It may have smeared all over the picture. The camera was in motion during the picture and the brain attempted to compensate for the lack of light, which should have held the shutter open for a touch longer, in turn, ruining the picture.

If you have other cameras, especially digitals, lying around the house, get them out and dust them off (be sure to keep the dust for the test!) and try the above steps again. You can also try this test with the camera setup we went through earlier. You should notice some differences in the pictures and will hopefully see how leaving the thinking to the brain of the camera is probably not the best idea during an investigation.

Over the past few years, I have been involved in several debates over dust orbs versus possible real orbs. Some investigators have attempted to come up with a system of discerning which is which, but I still believe they are all dust or other airborne particles. Here is a short list of orb-related lore I have heard over the years:

Different colored orbs:

Dust typically shows up white, though it's sometimes more of a brown, depending on the conditions where you live. Other colored orbs have shown up from time to time like red, green, blue, yellow, etc., but this does not mean they are paranormal. Pollen from plants and flowers will typically show up as yellow or green. Pocket lint from blue jeans will show up as blue and other weird airborne fibers will show up as red or some other color. Dust is not the only airborne particle. If you are outside there is no telling what could be floating by.

Partially hidden orbs that appear to be behind something:

This is probably the most disputed case for orbs but is really just an optical illusion. The orb is actually in front of the object, but due to the object's brightness, the orb light information is overshadowed by the object so it looks as though the orb is behind it. Another observation about this phenomenon is most of the orbs are typically on the left side of the object. This is due to most flashes usually being on the left side when the camera is facing away from the photographer. The flash intensity is much brighter on the left than on the right. This can also make them appear to be coming out of walls, doors, etc.

Orbs appearing to be in motion:

This phenomenon happens when the shutter hangs open for an extra millisecond or two. This is not unlike the test we did earlier where the artificial brain of the camera is making the decision that it needs more light, but the flash is almost out of light at that point. The particle did move, but the tail is created by the fading light of what is left of the flash.

Orb photos during the day:

This one is not too common but it does happened from time to time. But what you really need to ask yourself after you get one of these is, was the flash on? Just because it's daytime does not mean the flash is not bright enough to illuminate dust in front of the lens. Although the flash is no match for the sun when it comes to throwing extra light during the day, it will affect subjects within millimeters of the lens.

There are plenty of other questions you can ask yourself when it comes to orbs, things you may not have even of thought of or maybe you did. Here are some of my favorites:

How come you never see any orbs floating by during the evening news?

This is has to do with lighting. TV studios are well versed in lighting techniques. The main lights are high above the news anchors and many different light sources are used to remove shadows. The cameras are from five to 10 feet away from the anchors and typically located in a darkened area to stop streaks or other light problems from appearing in the lens of the camera. Mobile cameras like on-the-spot live shots use high-powered lights as well, but they are situated at the top of the camera and typically halfway back on the camera's body. Particles will get no illumination when they are right in front of the lens.

How come orbs only show up during flash pictures?

This was covered earlier but there is also a big misconception that the camera flash is bleeding into the light spectrum that the naked eye cannot see. Although this is true at night the camera flash has nowhere near the spectrum of the sun. This being the case, you would think orbs would be visible during the day time when taking non-flash pictures.

How come orbs never show up in studio pictures when they use a flash?

This is the same as a TV studio: lighting. A film studio is set up so no shadow will be cast when the picture is taken. In most cases several flashes are used and none of them are any where near the lens of the camera and most are diffused from being that harsh bright light of a point and shoot camera. There is no way a dust particle is going to get illuminated in front of the lens.

A sensitive said, Quick take a picture! and there was an orb near them, so that must be paranormal!

I believe this to be coincidence. I do believe in sensitives and work with them constantly during investigations, but just because an orb showed up when they said to take a picture does not make it paranormal. Since dust is everywhere, you have a good chance of picking up orbs at any given time no matter where you are, sensitive in tow or not, especially if you are in an old building that has not been cleaned in ages. I have plenty of pics where a sensitive said the same thing with no results of anything around them and even some where orbs appeared to be right above them or on their hand, etc. Just because a sensitive is feeling something does not mean the anomaly is going to show up in a picture or on a video.

My EMF meter was going crazy so I took some pictures and got orbs. That makes it paranormal!

EMF detectors can be affected by dust, believe it or not. Dust can get statically charged just as you yourself do in the winter when you walk across a carpet and touch a door knob. Most EMF detectors are quite sensitive and can pick up on the energy being emitted by dust particles in the air. If you want to see this phenomenon yourself, here is a simple experiment. On an extremely dry day or in the winter, comb your hair several times with one of those cheap black combs you can find at Wal-Mart for 99 cents. Have someone hide the comb while you are not watching, but make sure its not crushed under anything that could discharge the static, then use your EMF detector to find it. The EMF detector should start picking the comb up from a good distance before you get on top of it.

I took a picture of an orb. When it zoomed, it had a face. That makes it paranormal!

The mind is constantly working to recognize things we are familiar with and it can sometimes see familiar patterns in a garbled mess. Just like looking at the clouds and seeing the familiar shape of an animal or object floating by, the mind can pick up on familiar shapes and objects on your computer screen. This is called matrixing. We do this every day without even thinking about it. We will cover this in detail later.

Other contaminates can cause just as big a problem as dust. Objects attached to the camera and hanging in front of the lens, like camera straps and lens cover strings, can cause what looks like a vortex in the picture. Cat hair and sweater fuzzies can be a big nuisance and will not look like orbs but more like a streak or a zig-zagged, out-of-focus blur. Be careful to select garments that will not introduce strange fuzzies into the air. Flying insects can cause a blur or streak or an out-of-focus anomaly as well.

As you are working your investigation, be aware of your surroundings and look for things which could cause false positives. If you are unsure what some of these things might look like, use the test process from earlier and take some pictures. I have found the best place to test your camera is in a garage or barn setting, someplace which does not get cleaned often. The more you play, the more you will figure out.

So how do you stop the orb madness?

This is actually not too hard depending on the type of camera. The best method I have found is using an external slave flash along with your current flash on the camera.

Get yourself a slave flash with an optical trigger. What this does is use the red-eye reduction flash strobe from your existing camera to activate the slave at the same time. To make this work properly without causing

Slave Flash

contamination, mount the slave flash to your camera and get it set up and working according to the instructions. Once you have it working to your liking, take a piece of masking tape and place it over your camera's flash. What this will do is diffuse your on-board flash so it will not affect what's in front of the lens directly, but will let enough light out so it will fire the slave flash. Try the dust experiment from earlier to get the slave flash positioned so it does not cause contamination. Please keep in mind this is not 100 percent foolproof, but it should give good results.

One of the big misconceptions about digital cameras is a pixel on the CCD may not operate properly and cause what looks to be an orb in the photograph. Although this is not impossible, it's going to depend on how the pixels are wired into the CCD grid. Some are wired from top to bottom and some from side to side. If a pixel were to malfunction, it may affect the pixels above or below it, or the pixels on the left or right, but not both at the same time. It's almost impossible for a pixel malfunction to cause an orb. If anything, it may cause a straight line through the picture.

8. VIDEO

In the late '30s and early '40s a new technology emerged on to the consumer market that changed our lives and brought us instant entertainment. Television would forever change communication as we know it and probably make some people captive zombies in their own homes. In those days, televisions were created for one purpose: to receive television stations. At the time, video recorders, DVD players and video game consoles were only a pipe dream and all shows were performed live using motion picture film technology to record the events. But it would not be too long before television would expand into a total entertainment medium.

Television technology also brought us the video camera. Obviously, without some kind of input device, a television turns into a fifty-pound paperweight, but back in the day, video cameras were quite large and not very mobile so television studios were the only place they could effectively be used. Over the years, the cameras got progressively smaller, but unlike today's cameras which use charge coupled devices (CCDs) as discussed in the digital camera chapter, these older models used a technology called Orthicon tubes.

Orthicon Tube cameras, typically known as "tube cameras," bear a direct relation to television in their internal workings except they

Early Television Camera

collect light instead of producing it. Using the shoebox camera obscura from the earlier chapters as our model, if we cut out the area where the picture is projected and replace it with a thin piece of paper, we would be able to see the image projected onto the paper from outside the box through the paper. With this setup, we could trace the image to create a duplicate of the image on the back side of the paper. The camera tube basically does the same thing but is scanning a photoelectric plate with an electron beam instead of a piece of paper with a pencil. The image is scanned from one side to the other, all the way down the plate just like your eyes do as you read this book. Then it returns to the top to repeat the process for the next image.

The signal from the tube is fed through the cameras circuitry and eventually across the airwaves to your television, which is doing the exact same scanning except it's re-creating the image on your television screen line by line. A camera tube reads the photoelectric plate 30 times per second, so a television redraws the picture 30 times per second, following the same pattern, and, in turn, tricking the brain into seeing motion on the screen when in fact it's drawing static images which change ever so slightly every

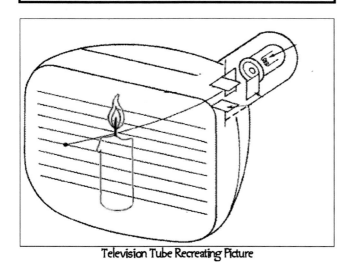

Television Tube Recreating Picture

Early Portable Video Camera

1/30th of a second.

Consumer video camera models started hitting the market in the late '60s, but like any new technology, they were quite expensive and definitely not the same quality as those in TV studios. The first consumer models were quite large, with the camera and recorder being separate. These units were great for daylight recording but nighttime was a different story. For the most part, these units were worthless at night even inside with all the lights on. By the late '70s and early '80s, tube cameras would advance to the point that they could work in lower light settings, but there would never be a Night Shot version available to the general public like today.

Tubes were the primary device for capturing video in professional and consumer cameras up until the mid-'80s, when CCD technology reached the point where it matched up to the basic tube quality. These CCDs were much smaller, lighter and consumed far less power, which made them a great replacement for tubes. The '80s would also bring us a camera and recorder built into one device called a camcorder. Several camcorders were modeled around tube technology, but they quickly moved to CCD as it became available.

The CCD technology of a

First Tube Camcorder

video camera works exactly the same way as a digital photography camera except the video camera is constantly taking pictures. Like tube cameras, CCD video cameras take 30 pictures per second, which is the standard for the United States. Other countries may be different. Once the image is captured, it is written to some kind of recording media like a VHS recorder or DVD writer.

Tube cameras have pretty much disappeared from the market, though some types are still produced for the medical industry and other specialized uses. Some professional videographers still swear by tube technology and continue to use it for various productions, but for the most part, CCDs are now the standard. CCDs also brought us the Night Shot technology that Sony boasts in many of their camcorders, but tube cameras were not really lacking in this ability. In fact, paranormal investigators may find value in this antiquated technology.

Both cameras need light to operate properly, although CCD is much more efficient and requires far less light to actually generate a picture. The amount of light required is rated in "lux." One lux is basically equivalent to the amount of light one regular wax candle creates. So if a camera is rated at 35 lux, you would need the same amount of light as having 35 candles lit to get a enough light to generate a picture. Most cameras today only require one lux or less to generate a picture, although the more light, the better the picture is going to be.

From the digital camera chapter, we know a typical CCD can see about 350nm to 1200nm of the electromagnetic spectrum. This basic fact makes them well suited for picking up near infrared light and some ultraviolet light. Tubes are a bit different and can see about 250nm to 900nm of the electromagnetic spectrum. Although they cannot see as far into the infrared range as a CCD, they can see more of the ultraviolet range. One of the downfalls of a tube camera is it requires more light to see the same scene as a CCD, so using a tube camera for infrared recording is possible if you have enough infrared light available. The same goes for UV light.

Theories of spirits captured with video cameras are basically the same as those for digital and film photography but also include another theory, which specifically deals with orthicon tube cameras. These cameras require a good bit of power and like a TV, incorporate circuitry that will produce thousands of volts of power to drive the tube, which may attract spirits to the device. Unlike CCDs, which are reading the light by using little sensors, tube cameras are bombarding a metal plate with electrons that can be severely effected by electromagnetic disturbances, specifically if it's close enough to the camera. If a spirit attracted to the device gets close enough to affect the electrons, the disturbance could be recorded and possibly presented as a paranormal event.

Finding the right camcorder for paranormal investigations can be just as big a pain as selecting a digital still camera. There are plenty of manufacturers boasting many different features and video formats that can get confusing. As with the digital still cameras, the size and structure of the case can be just as important as the features and ease of use. Video quality is always a concern whether you are ghost hunting or general recording. Buying the $2.50 Betamax camera from eBay is probably not the best idea, but you would definitely be the topic of many discussions after your first investigation.

There are two main types of camcorder: digital and analog. Most camcorders manufactured today are digital, but there are still a couple of analog recorders on the market that are rather inexpensive but may not be worth the money as the recording medium may disappear from the market in the next few years.

Analog video technology has been around since television was first created and uses electronic pulses, which are broken down to electromagnetic pulses and recorded to the tape. During playback, the electronic pulses are read from the tape to reproduce the picture and sound. The problem with this technology is the information can fade over time and an otherwise perfect recording today can look and sound pretty bad after 10 years of storage. Most analog video tapes will hold the recorded content for about 30 years before the information completely fades. Analog media includes Beta, VHS, VHS-C, S-VHS, S-VHS-C, 8mm and Hi-8 videotapes.

Digital video technology uses binary data, which is 1s and 0s produced by a microprocessor then electromagnetically recorded to the tape. Although these can also fade over time, since it's a simple 1 or 0, the quality pretty much stays intact as long as the 1s and 0s can be read from the tape. Digital provides a far better picture and clearer sound than analog technology and also offers more recording media options. Digital includes Digital 8, Mini-DV and Micro-DV tapes as well as DVD and hard drive recorders. Some digital camcorders also offer the ability to record to removable computer memory like SD Cards and Memory Sticks.

Selecting a format isn't crucial for ghost hunting, so it's more of a personal preference. Be sure you can manually set the focus, iris and shutter speed. Also, look for a camera with infrared capabilities. If you are looking at the newer hard drive style cameras, be aware most only hold about 30 to 40 hours or so of high-quality video. In order to back up the data on the drive, you will need a recent computer with plenty of hard drive space or a DVD recorder to archive the files.

Using your camcorder in the field is not much different than using it for everyday recording, but just like a digital still camera, using the manual settings can help a great deal. Lower light settings seem to wreak havoc on the internal brain of the camera and can cause the focus to go haywire at any given time, especially when using infrared mode. Solving this is as simple as putting the camera into manual focus mode and zooming to the widest setting on the camera. This also helps with battery life as the camera does not need to constantly adjust the focus.

Infrared mode can also drain the battery faster than normal operation. Using an external IR light source can help with this and also brighten the area a bit since most IR illuminators produce more light than the one built into the camcorder. Sony offers a nice illuminator that connects to the bottom of the camera using the tripod mount. This will ensure the light is always pointed in the same direction as the lens.

Camcorders are great not only for documenting your experiences while investigating, but also for offering unattended operation. Many investigators have that found leaving the camera in a quiet location for a few hours can yield interesting results. It is best to use a tripod with the camera in the manual settings we talked about earlier. Make sure the tripod is on sturdy ground and all knobs and levers are tight.

Looking for evidence after the investigation on a camcorder can be the most tedious task you will undertake. Watching two hours of video of the same scene can get pretty monotonous, but if you take it in 20-minute increments, you won't find yourself clawing at the walls waiting for something that may never happen. A preliminary scan through the footage using fast forward can also help find any major events quickly. Using this method is actually very effective since you are more apt to be paying attention during the fast forward than while watching the tape in

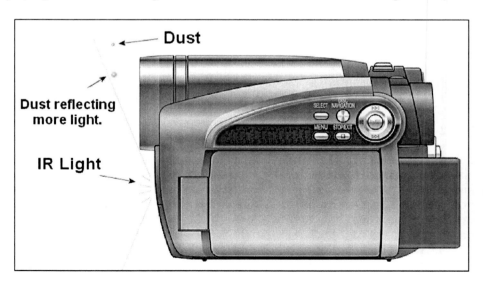

real time.

Camcorders can be plagued by the same contamination issues as a digital still or film camera, but the problem can actually be worse since the contamination will most likely be moving. Dust is probably the biggest contributing factor, as it may produce an optical illusion that something is moving either toward or away from the camera. For example, if the IR light is below the lens, dust falling in front of the lens will reflect more light as it gets closer to the source, producing the illusion that it is getting bigger when, in fact, it's just reflecting more light. The opposite will happen if the IR light is above the lens.

The best solution for this problem is to use an external IR illuminator placed farther away from the lens and deactivating the onboard IR light.

Exploring waves beyond near infrared requires a special video camera called a thermal imager. This type of camera actually picks up on heat that is radiating from the body and other objects instead of using light like a normal camera.

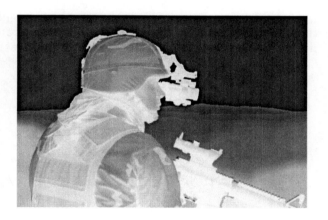

Thermal imaging devices were invented back in the late '50s and early '60s for military use using specialized camera tubes that eventually found their way into other industries in the early '70s when they were declassified for public use. Like camcorders, many newer thermal imagers use CCD technology instead of tubes, which make them much lighter and easier to work with. Although the technology has been around for many years, it's still expensive and few paranormal groups are able to afford them.

There are two basic types of thermal imagers: cooled and uncooled. Older models were generally cooled due to using camera tubes that generate heat.

Thermal Imager

This made it difficult for the device to produce a usable image when the internal electronics raised the temperature above the ambient air temperature. To solve this problem, liquid nitrogen was used to cool the circuitry and the tube to keep it at a constant temperature.

Uncooled imagers use the newer CCD technology, which does not require anywhere near the power to operate as a tube and does not generate nearly as much heat during operation. The temperature of the device may rise during operation. This version typically produces a good image although the cooled version is much more sensitive.

It is speculated that spirits will affect the ambient air temperature surrounding them and may be able to be viewed through a thermal imaging device. These devices sense heat instead of light and they can be plagued with various issues such as reflections off of smooth surfaces and heat signatures transferred to an inanimate object, which take a little time to cool back to the ambient temperature. As this technology advances and becomes more affordable, this device will definitely be very useful during paranormal investigations.

9. MATRIXING

From the time we are born until the time we die our brains are processing and visually memorizing various shapes and images that we associate with everyday objects and people. This basic function of the brain helps with activities like seeing a stop sign from several hundred feet away and recognizing what it is before you can read it or see its detail to seeing familiar shapes in the clouds. It's probably something we take for granted, but it can cause grief while looking for evidence of the paranormal.

Matrixing is basically the brain's natural ability to pick familiar shapes out of a completely unrecognizable scene. This is one of those areas that allow a person to be presented with a possible anomaly, typically in the form of a picture that may look like solid evidence, but in reality is just a lighting problem or flaw.

The best form of matrixing I have come across so far is the picture on the previous page. This is truly a marvel and can show how the eyes and the brain can be tricked into seeing something which is not reality. In this case, the image appears to be moving, when obviously it is stationary. If you look at it in different places, you can stop it from appearing to move in others, but for the most part, it always looks like it's moving. This is obviously an exaggerated version of what you might pick out of a photograph, but it can definitely show the power of the brain and how easily we can be tricked by patterns.

Matrixing can materialize in many different forms, but the most popular would probably be a human face. One of the most famous faces to appear in many different places is, of course, Jesus, but the Virgin Mary has also shown up quite a bit. How does one explain the sudden appearance of a face in everyday products such as bread, or in my case, a can of clam chowder? I am not really sure what to make of this phenomenon, but if I traced my family tree back far enough, I am willing to bet I could find someone who resembles the goop in this can. Is this a form of communication from a departed loved one or just a

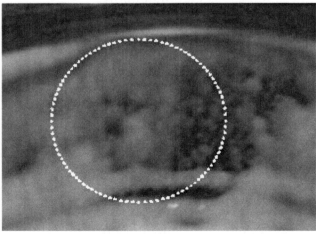

good optical illusion?

Some people are better at picking up on this phenomenon than others, although most people will see it once it is pointed out to them. Whether it's a face in a soup can or the Virgin Mary on a piece of toast, without some other form of evidence, it's really hard to claim this to be anything other than being in the right place at the right time to see a good illusion. Many psychics and spiritualists believe otherwise and regard this as a possible sign or attempted communication from the other side.

Seeing things out of the corner of your eye can also be a form of matrixing, especially in a dark room or during a nighttime investigation. Glimpses of cobwebs hanging or bugs flying within inches of our eyes can be difficult to discern and our brains may interpret them as something else. There is also a physiological claim to this in the form of "eye floaters."

Eye floaters are tiny clumps of gel or cells inside the vitreous, the clear jelly-like fluid which fills the inside of the eye. For the most part, they are transparent but they can sometimes be seen if you stare at a solid-colored wall or blank piece of paper.

Floaters may look like specks, strands, webs or other shapes. Actually, what you are seeing are the shadows of floaters cast on the retina, the light-sensitive part of the eye. These can actually become a problem at night since the pupil will open up as far as it can to let in more light. Since your peripheral vision is more sensitive to light than your regular vision, floaters passing through your peripheral vision can leave the illusion that something or someone just went darting by when, in fact, a floater cast a quick shadow on the retina. Unfortunately peripheral vision is not very good with detail, so recognizing a particular person or shape is not very easy.

10. ITC

Instrumental Transcommunication (ITC) is defined as using a technical means to receive meaningful messages from the spirit world. If you have been involved in paranormal investigations for any period of time, you have probably attempted or at least heard of EVP (Electronic Voice Phenomenon), which is where ITC originated. ITC encompasses a myriad of techniques to attempt communication with the spirit world through the use of television, radio, audio recorders, telephones and even computers.

Communicating with the other side through electronic devices is not a new idea or concept and it has been attempted since audio devices were invented. The first known recording of spirit voices was captured by ethnologist Waldemar Bogoras in 1901 during a trip to Siberia to visit a shaman of the Tchouktchi tribe. Using a gramophone recorder, like the one pictured earlier, during a spirit conjuring ritual, Bogoras recorded spirit voices in English and Russian which could be heard during playback. These are usually considered to be the first communication using an electrical recording device.

Unfortunately, this is not totally true. It may be the first recording of spirit voices, but the gramophone recorders of the time were completely mechanical. These devices used a needle connected to a large megaphone to etch sound vibrations onto a record, which was rotated using a hand crank. Electric gramophones would not be available until about 1925.

Various inventors in the early 1900s had theories on life after death and some even attempted to create devices for communication. Thomas Edison was supposedly working on a secret machine that would allow direct communication with the spirit world through the use of electronics. Unfortunately, Edison entered the spirit world himself before the machine could be completed. Several eyewitnesses reportedly had the rare pleasure of seeing the machine, but the plans and the machine itself disappeared after Edison's death. It is speculated that his colleagues destroyed the machine and the plans before anyone could find them.

Although this information appeared in many newspapers across the country and several magazines, the Edison Museum maintains this device never existed and the whole story was concocted from Edison's interview with *Scientific American* magazine on Oct 30, 1920, where he was quoted as saying, "Nobody knows whether our personalities pass on to another existence or sphere, but it is possible to construct an apparatus which will be so delicate that if there are personalities in another existence or sphere who wish to get in touch with us in this existence or sphere, this apparatus will at least give them a better opportunity to express themselves than the tilting tables and raps and Ouija boards and mediums and the other crude methods now purported to be the only means of communication."

> **EDISON AT WORK ON SPIRIT DEVICE**
>
> Seeks to Communicate With Dead by Use of Electrical Instruments.
>
> New York, Sept. 30.—To communicate with the spirit world by means of delicate electrical instruments is now the goal of Thomas A. Edison.
> Back of the wizard's devotion to his new task is the thought that if communication is ever really established with the personalities which may persist after death, it will come through science and hard work and not through mysticism or such clumsy contraptions as the ouija board.

ITC did not really take off until 1959 when Friedrich Juergenson captured a disembodied voice saying something about "bird voices in the night" while he was recording bird songs in the park with a tape recorder. Juergenson continued to experiment, recording hundreds of voices, which prompted him to published a book in 1964 titled "Voices from the Universe," followed by a second book called "Radio Contact with the Dead."

In 1967, Latvian psychologist Dr Konstantin Raudive read the German translated version of Juergenson's second book, which inspired him to visit Juergenson and learn his recording techniques. Soon Raudive was recording his own spirit voices. He published a book in 1968 called "The Inaudible Becomes Audible," where he discussed the more than 72,000 voices he recorded. Some of Raudive's original recordings can be found on the Internet by searching for "Raudive audio."

Many forms of ITC have been attempted with success over the years although no one device or method has been known to produce constant results or even to work for every person. Exceptional patience and an open mind are your greatest assets when attempting contact with the other side.

11. SOUND RECORDING

Recording audible vibrations is a simple process that was invented in the mid-1800s even before Thomas Edison's phonograph. The actual process of recording sound was accomplished in 1857 by French printer and librarian Leon Scott using an invention he called the phonoautograph.

This simple apparatus used a small bristle connected to a large diaphragm to inscribe an image onto a rotating cylinder. Though the device did not have the ability to play back the sound, it was useful for the scientific exploration of sound waves.

Unlike light waves and radio waves, sound waves are not electromagnetic in nature but share a common trait being rated by frequency which for sound is cycles per second called hertz. If you string a rubber band across your hand and pluck the longest part between your thumb and pinky you will be creating sound. If you wanted to rate this sound in hertz, you would need to count the amount of times the rubber band vibrates back and fourth in one second.

You can make the pitch higher by stretching your thumb and pinky farther apart, which will make the rubber band vibrate faster or slower by bringing them closer together. The intensity at which you pluck the rubber band can change its volume. The harder you pluck, the louder it will get but it will maintain the same pitch as long as the pinky and thumb stay the same distance apart.

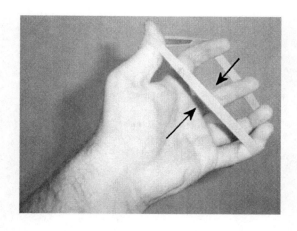

Sound travels by passing vibration from the source to airborne particles, which basically hand the vibration across to other airborne particles and eventually make it to your ear. The sound weakens between each handoff so eventually the sound can no longer be heard. The bigger the vibration, the louder the sound will be and the farther it will travel.

The human hearing range is from about 20 Hz (Hertz) to about 20,000 Hz though most adults can only hear sounds from 300 Hz to 13,000 Hz, depending on their age. As we grow older, most of us start losing the ability to hear higher pitched sounds. The typical human voice produces sound from 300 Hz to about 3,500 Hz during normal conversation.

Twenty years after the phonoautograph was invented, Thomas Edison introduced the phonograph. Using a similar concept as the phonoautograph, the phonograph incorporated a large horn connected to a needle with a small cylinder that was covered with tin foil, wax or lead (called a "phonograph cylinder") as the recording medium. As the cylinder was rotated under the needle, sound vibrations entering the horn would vibrate the needle, in turn etching the sound onto the cylinder. The process for carrying sound was not much different than the typical paper cup telephone we played with as children.

A similar device called the gramophone was introduced in 1878 by Emile Berliner, a German immigrant. Berliner's

Edison Phonograph Mechanical Drive

device used discs called "gramophone records" instead of cylinders and is the base invention that brought us LP records. Although modern day record players are called phonographs, they are actually gramophones. Edison struggled for many years attempting to sell his phonograph cylinders against the gramophone records, but finally gave in and created his own version of the gramophone in the late 1920s.

These devices were able to record and reproduce sound fairly well; however, neither used any electronic parts to do so. The large horn was used as an amplifier to capture the sound that funneled down to the needle during recording and worked in reverse by amplifying the sound from the needle during playback. The cylinder or record was kept at a constant speed using a similar concept as that found in wind-up clocks. It was quite a sophisticated process for the late 1800s.

Vlademar Poulsens Telegraphone

Although gramophones and phonographs are given most of the credit for giving us dictation abilities and music reproduction, another device, called the telegraphone, invented by Valdemar Poulsen in Denmark, was introduced around the same time. This device was initially designed to record telegraph conversations and eventually telephone conversations, which the phonograph or gramophone couldn't do. Unlike the other devices, this machine was the first to use an electromagnetic process to record written information using a steel wire wrapped around a large drum.

Electronic valve tube amplification, microphones and motor drives were added in the mid-1920s after World War I, making these machines easier to use and producing far better sound quality. As electronic technology advanced, newer devices started to emerge on the market with far better quality than the old gramophone records and phonograph cylinders.

Wire Recorder

The telegraphone would eventually turn into the wire recorder, which resembles the present day reel-to-reel systems. Instead of tape, a steel wire is wrapped between two spools and drawn across an electromagnetic head for recording and playback.

In the early '30s, different devices emerged onto the market using special iron oxide powder coated onto a plastic tape and an electromagnetic head. This device was called the magnetophone. Like the Betamax vs. VHS war of the early '80s, both the magnetophone and wire recorders were battling for corporate and consumer attention. The wire recorder promised a better quality machine at the time, but the magnetophone eventually won.

Technology of the early '50s brought us the

Magnetophone

cheaper and more efficient transistor to replace the old valve tubes, which required a good bit of power to operate. Transistors eventually made it possible to make radios, tape recorders and even televisions much smaller and more portable.

The magnetophone is actually a reel-to-reel tape recorder, but the term "reel-to-reel" did not come about until the early '60s when cassette type tapes started hitting the market. All subsequent magnetic tape recorders were modeled after the magnetophone, so to discern the difference between the cartridge-style machines and the open reel-type machines, the name "reel-to-reel" was adopted.

Different styles of cassette tapes were produced from the late '50s to the late '60s, but the format that gained the most popularity and eventually would become the standard was the compact cassette. The runner-up being the microcassette. Both of these were created about the same time, but compact cassettes would eventually become the favorite.

All magnetophone-style tape recorders work on the same principle, using an electromagnetic head to record and read the information from the tape. This process is electrical, but it's not too terribly different than the process used to record sound onto a gramophone record.

Electromagnetic Record Head

The microphone receives a sound in the form of vibrations, just like your ears. These vibrations are turned into electrical pulses and fed to the amplifier, which strengthens the pulses. The resultant signal is

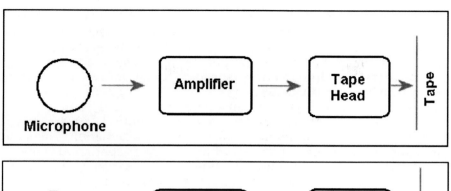

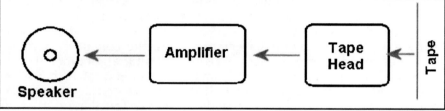

fed to the tape head, which aligns the electrons on the passing tape magnetically, in turn storing the information. This is basically the electronic version of etching vibrations onto a wax cylinder.

The entire process is reversed during playback. The tape head reads the signals from the passing tape, which are sent to the amplifier to enhance the volume. After amplification, the signals are sent to the speaker, which vibrates to reproduce the sounds much like your vocal cords vibrate to create sound when you speak.

The tape is drawn past the head at a constant rate, which dictates the quality of the recording. The faster the tape passes over the head, the better the quality is going to be. A standard compact cassette recorder draws the tape past the head at 4.75 centimeters per second (cps). At this rate, the entire sound spectrum is able to be recorded

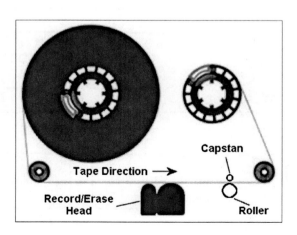

and played back with good quality.

The smaller microcassette recorder is designed to record the human voice and draws the tape past the head at 2.4 centimeters per second (cps). This is almost half the sound quality when compared to the larger compact cassette. Although this speed is good for voice recording, music quality is seriously degraded.

Phonographs, gramophones, magnetophones and all forms of consumer audio recording devices created up until 1980 are analog. Analog is defined as continuous transmission of information. This is not unlike the gramophone record, which is one continuous line of information from start to finish. As humans, we perceive the world in analog; everything we see and hear is a continuous transmission of information to our senses.

In the early '80s a new media emerged onto the consumer market in the form of digital compact discs (CDs) and digital audiotapes (DAT). It seemed like a new concept at the time, however digital technology is almost as old as wire recorders. Alec H. Reeves invented digital encoding (pulse-code modulation) in 1937 while working for the International Telephone and Telegraph Co. in France. Bell Labs used this new technology for private phone calls through a telephone encryption system during World War II.

Digital is defined as binary information in the form of a 1 or 0. Make sense? Yeah, I didn't think so either. In a nutshell, digital is like Morse code for a computer processor, which allows the information to be encoded and decoded from a magnetic tape or disc in the form of 1s and 0s. Each 1 or 0 is generically known at a bit.

The digital recording process starts just like an analog recorder using

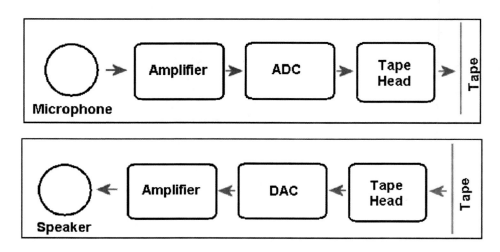

a microphone to convert the sound to electrical pulses and then amplifying the pulses. The constant analog pulses are converted to digital information using an analog to digital converter" (ADC). The information is then recorded to the tape using an electromagnetic head.

To play back, the digital information is read from the tape using an electromagnetic head. Then it is sent through a digital to analog converter (DAC) to reproduce the original analog electric pulses, which are sent through the amplifier to vibrate the speaker.

Digital information is a bit more complex when it comes to quality. The speed of the tape passing the head plays a small role, but the bulk of the sound quality is made up by how much information can be converted through the ADC process during recording and the DAC process during playback in one second. This is known as bit rate.

The bit rate for perfect quality sound is 1,411,000 bits per second (bps), which is the sound quality of a store-bought music CD and the base bit rate every digital device is compared against. In most cases, at least 1,000 bits are being written or read each second so many manufacturers use the next higher measurement of kilobits per second (kbs).

Using this information, let's take a look at one of the most popular recorders for paranormal research, the Sony B100, and compare it to the perfect quality sound of a CD. A typical store-bought audio CD has a bit rate of 1,411 kbs. Keep in mind that most digital recorders have different quality modes, so we will deal with the highest quality mode of the recorder as our comparison. Sony rates the High Quality (HQ) mode bit rate of this device as being 19.2 kbs. As you can see, this device is nowhere near the quality of a professionally recorded CD.

This is really a crude method of comparing these devices and is really comparing apples to oranges when it comes down to it. There are several other factors involved that dictate the overall quality of each digital device, which are beyond the scope of this book.

If you are interested in learning more about how digital and analog technologies compare, there is plenty of information to be found on the Internet.

12. ELECTRONIC VOICE PHENOMENON

Electronic Voice Phenomenon is spirit voices caught on some kind of electronic audio recording device that is not heard by the human ear during the recording. The spirit voices are only audible during playback. EVP is probably one of the creepiest forms of evidence you can acquire during an investigation, especially when a disembodied voice answers a question with an intelligent response. Even the biggest skeptics can be left with more questions than answers and EVP may even make believers out of them.

This phenomenon is sometimes considered a form of ITC and in other cases is considered a stepping stone for what ITC has become over the years. Whatever your opinion, EVP is definitely one of the most popular techniques used during a paranormal investigation and does not require any specialized equipment. A simple audio recorder, a quiet room and a little patience is all that is needed. Just about any type of audio recorder can be used and the choice is left more to personal preference than it is to any proven technique or recorder type.

Shopping for a recorder doesn't need to be difficult, but you may want to decide what kind of recorder you are considering before heading out to the local electronics store. Not all recorders are created equal, but all pretty much have the ability to capture EVP.

If you are considering an analog cassette recorder, you will definitely need to have an external microphone. All tape recorders use internal motors to move the tape, the on-board microphone will pick up this sound during recording, which can hinder your ability to actually hear any EVP and is definitely a nuisance during playback. Most present day recorders have an external microphone connection, but check to be sure since this is not always true.

Analog tape recorders come in two basic flavors: compact cassette and microcassette. As discussed in the sound recording chapter, compact cassette recorders will typically give you the best audio quality. Most people are not very happy with their overall size and weight and will most likely lean towards the microcassette recorder. If you are considering a microrecorder, keep in mind the sound quality is not the same.

Compact and microcassette recorders use the same mechanisms. The difference in sound quality has to do with the speed of the tape and not the construction or electronics of the device. As stated earlier, tape from a standard compact cassette passes by the head at 4.75cps. The reason this speed was selected is to effectively cover the frequency range of human hearing for use with voice and music recording. Any slower and the sound quality starts to degrade.

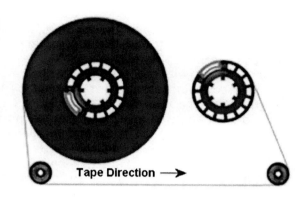
Tape Direction ⟶

Micro recorders were not really designed for full-range audio recording. Instead, they were designed around the range of the human voice and the miniature size of the tape itself. Since the cassette is much smaller, it holds about a third the recording time of a full-size tape. To get the same recording time as its big brother, the tape of a microcassette is passed by the head at 2.4cps. Being almost half the speed of the compact cassette degrades the sound quality to almost half as well.

In basic terms, micro recorders were designed for lectures and word dictation and not to record the sound of a full orchestra. Obviously, you are not buying the recorder to sneak into the next Metallica concert and record yourself a bootleg, but being half the quality can sometimes cause an issue during an investigation.

Micro recorders will work just fine for EVP; the problem is when something happens in the background during recording that is out of its frequency range but completely audible to the human ear. This can sometimes cause strange results that could be mistaken for an EVP. However, the sound was actually the high-pitched creaking of a floor or the wide-open throttle of a passing motorcycle.

Many micro recorders also offer a slower tape speed setting at 1.2cps. If you really want to torture yourself with audio garbage which sounds like possible EVPs, this is the setting to play with. I would not even suggest this setting for general voice recording. The slower tape setting is pretty much worthless and is not recommended for EVP research.

Both compact and microcassette recorders typically offer voice activation, which is also known as VOX. This convenient feature starts and stops the tape when the recorder's microphone picks up a sound. Although this may sound like a great idea, it does not work well when trying to capture spirit voices.

There are actually two problems with the voice activation feature. The first is, a communicating spirit may not be loud enough to activate the recorder. Imagine your fellow investigators getting a Dennis Miller-style rant from an agitated spirit and you get nothing because your recorder doesn't start the tape. The second problem is, it takes half a second for the recorder to respond and kick on the motor to start recording. You could miss half or all of an EVP because the recorder was not quick enough to start. It is best to deactivate this feature during an investigation.

Another annoying feature that can be found on both kinds of recorders is auto level control. This basically tries to keep the audio level of the recording at a constant volume at all times. Any loud noises like a door slamming or someone talking right next to the microphone will activate this feature, which will lower the incoming audio level until the loud noise goes away, then it will turn it back up. This can cause you to miss portions of an EVP because the microphone level dropped.

Analog recorders have been used for EVP research for years; you really can't go wrong no matter which one you select. Don't forget to use high-quality tapes to get the best sound possible.

Trying to find the right digital recorder can leave you praying for the nice guys in the white coats to show up and take you on a pleasant retreat. There are so many different manufacturers and models to choose from that it's difficult to even figure out where to start. I would personally stick with the bigger name brands, which have put a ton of research into their products and pride themselves on sound quality and ease of use over manufacturing the device as cheaply as possible.

Recording quality is going to be your biggest concern, which most name brand recorders do quite well, but remember the discussion on comparing the Sony B100 to the audio quality of a compact disc. Most digital voice recorders are not going to get anywhere near perfect sound quality, but spending some time checking out the bit rate of the recorders can help in making a good purchase decision.

If you are lucky enough to go to a store with working display models, try making some sample recordings on the highest quality setting and listen to the playback carefully. Some recorders may have digital garbage in the background that can produce false EVPs. Obviously you will want to avoid this issue as much as possible.

Digital voice recorders generally lack internal moving parts, so an external microphone is not really necessary but is always a nice feature and usually preferable during an investigation. Rechargeable battery packs can be convenient but can be a major problem as well. Although the device may claim the battery will last for 10 hours, spirits have been known to drain battery power in a very short period of time. Unless you are planning on buying extra battery packs, which normally cost an arm and a leg, it is best to get a recorder that will run on typical store-bought alkaline batteries.

Voice activation can also be found on most of these recorders, but unlike cassette recorders, which require a motor to start moving before it can actually do any recording, digital recorders can start and stop almost instantaneously. You may still have an issue with the recorder not hearing the sound to get it started, although this is not as big a problem for digital recorders. Cassette recorders must compensate for the sound of the motor drive and limit the voice activation sensitivity or the recorder would never shut off once it gets started.

Select a recorder you feel comfortable with and keep in mind many EVP sessions are conducted in the dark, so you will want to be able to operate your device pretty easily without really being able to see what you are doing. Most newer recorders incorporate some kind of light to signal when the device is in record mode or at least operating to some capacity. Get to know your recorder before you take it out in the field.

Preparing for an EVP excursion is not too terribly difficult, but it does require some common sense. Being as quiet as possible during a session is probably one of the hardest things to do and can have more to do with what you are wearing and how you are holding your recorder than it does with not actually talking. Your recorder is a lot more sensitive than you may believe and little faint noises near your recorder that you can barely hear can sound like a nuclear explosion to the recorder's microphone.

Don't wear clothing made of plastics that make noise when you walk or move; this can easily be misconstrued as an attempted communication. Although you may hear the sound loud and clear on your recorder, others in your group may hear it faintly and not know what it is. Clothing made of cloth is a better choice and typically does not make any noise.

Dangling jewelry can be an issue as well. Bracelets that clang when you move your wrist can make a completely different sound to your recorder since the frequency is typically too high for the recorder to hear properly. Loose watches can make a similar sound. It is best to leave these at home or remove them before starting your investigation; you don't want to contaminate your evidence.

Holding your recorder during the sessions is not recommended and is probably the biggest cause of contamination when attempting EVP. To give you an idea of the problem, take your index finger and lightly run it along your outer ear near the canal opening. You will hear what sounds like static or white noise. No one else will be able to hear you doing this even if they are standing right next to you. The sound is very faint, but since the outer ear is attached to the inner ear and eardrum, you can hear the vibrations caused by your finger loud and clear.

Recorders have this exact same issue. The microphone works much like your eardrum and is attached to the case of the recorder. Manufacturers put a lot of time into isolating the microphone as best they can, but it's almost impossible to eliminate case vibrations completely. Holding the recorder in your hand can cause all kind of crazy noises that are not audible during the initial session but will be heard during playback. External microphones can help with this issue, but holding the mic will cause the exact same problem through the microphone case.

The best way to eliminate case vibrations is to put the recorder and external mic on a towel or some other soft cloth and sit it on a sturdy surface. If you are using an analog recorder, put the recorder behind the microphone so you do not hear the motor drive inside the recorder. This will stop external vibrations from reaching the recorder during the session.

If you are outside and must hold the recorder, use an open palm with a soft cloth. Don't grip the recorder; this will make the vibrations even louder. Just let it rest in your palm. Using an external clip microphone attached to the back of your collar is a great way to eliminate extraneous noise and even your own breathing on the mic by accident.

If you must talk during the session, just talk normally, don't whisper. Your recorder may be close enough to actually capture the whisper correctly, but someone else 10 feet away may also capture it but not be able to tell what it is during playback. Believe it or not, recorders can be that sensitive. It is best just to talk normally so everyone knows it was you and can recognize your voice.

There really is no tried and true technique to capturing EVP; you have to try them all to see what works best for you. You need to keep in mind that spirits were people at one time, and as with the general public, they may not want to talk to you and may be drawn to someone else in your group. The psychology of spirit communication is kind of a new field and very little is known about what causes them to want to interact with certain people and not with others, but it's most likely the same psychology which attracts the living to certain people and not to others.

Before starting your investigation, come up with a game plan as to how and when you want to attempt EVP. This really needs to be a group exercise so everyone is aware of what is going on and knows to be quiet and not create any kind of noise. If you are going to ask questions, it may be a good idea to write them down and have them handy.

Doing a little research about the area or building you are investigating can give you some ideas on what kind of questions you may get a response to. Asking what kind of car they drove on the Gettysburg Battlefield will most likely yield no results, but asking for the name of their commanding officer might produce a response.

There really is no right or wrong way. Some folks find it easier to act like they are holding a direct one-on-one conversation with the spirit. Introducing yourself is a great way to start off a normal conversation. Spirits may appreciate this as well. Speak loud and clear and be sure to wait in between your questions for a response.

Another popular technique is to leave your recorders unattended in a quiet location. Be sure everyone knows where the recorders are and not to enter the room or make any loud noises or talk near the room where the recorders are located. Most recorders are quite sensitive and will pick up noises outside a closed room pretty easily. Be sure to announce your departure from and arrival to the room.

EVP is not limited to audio tape and digital voice recorders and can be picked up on other devices like camcorders and anything which records sound as part of its process. If you are not checking your camcorder for EVPs, you may be missing out on some possible evidence. Take your time and review the video and audio separately so you don't miss anything.

THEORIES

How and why EVP is captured is an explanation still awaiting an absolute answer. Many different theories have been scrutinized over the years with no way to really prove them. As technology moves forward on audio recording devices, we can look back at what has worked over the years and come up with some pretty good ideas of how this may or may not work.

One of the oldest theories, which has pretty much been shot down by new and emerging digital technology, is the spirit's voice is somehow magnetically imprinting itself onto the tape during the recording process. This would explain why you couldn't hear the words while they were being spoken. Unfortunately, if this were true, digital recorders, which do not use magnetic tape, would not capture EVPs.

Obviously, it's possible the EVPs are entering through the microphone just like human speech. It is speculated the spirit voices are above the human hearing range, which is why we do not hear them. The main issue with this theory is that most recorders used today are really only designed to record the human voice and cannot hear above or below the human hearing range.

I've done some in-lab testing of this theory using ultrasonic sounds (noises above the human hearing range) and some very high-pitched noises, which are within the human hearing range. I used the Sony B100 digital, Olympus WS-300M digital, Olympus compact and microcassette analog recorders for this test. For high-pitched hearing range noises, the Olympus WS-300M digital did the best job and could hear noises up to 20,000Hz with the B100 coming in a close second at 19,000Hz. Although these are just slightly within the human hearing range, I could barely hear them. The compact cassette did okay at about 18,000Hz and the microcassette did the worst at 13,000Hz.

During the initial tests with ultrasonic noises, my cats became interested, staring intently at my computer, apparently able to hear what I could not. I used a tweeter connected to my computer's sound card, which was rated up to 32,000Hz. Please note the cats were not meant to be part of this test and I did not take into consideration their hearing range until after I started. Cats can hear three times the range humans can. Luckily the volume was about 3, so it just interested them and did not hurt them. When playing back the recordings on all the recorders it did not have any effect on the cats, even using the tweeter hooked to the earphone jack. Although this does not totally prove these recorders did not pick up the sound, I didn't hear anything and I presumed the cats didn't either, since they were calm and seemed completely uninterested in what I was doing.

To verify my computer was putting out 28,000Hz, I used an ultrasonic down converter, which is a device for listening to bat sounds. This takes noises above our hearing range and makes them audible. I could definitely hear the sound coming from my computer through this device. I then played the recordings back and could not hear anything through the bat listener. I personally have my doubts about voice recorders hearing ultrasonic sounds, but it's not impossible.

Fellow researcher Rick Fisher from the Paranormal Society of Pennsylvania has also been experimenting with the ultrasonic bat detector that I used in the previous experiment. With the bat detector connected to his Sony B100 recorder, he has been successful in recording several voices using this method. Strange as it may sound, some of these voices were recorded from inside his microwave oven.

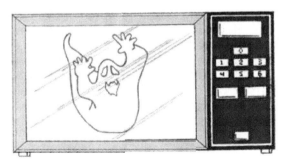

This is not to say there is a ghost stuck in Rick's microware, but the electrical properties of the microwave cabinet itself are a great place to filter out stray radio stations and other possible electrical noise. Rick has had similar results with a recorder by itself. If you decide to give this a try, do not turn on the microwave! You may destroy your recorder as well as the microwave and possibly start a fire. Unplug the microwave from the outlet to ensure your safety.

Rick and I have done many different experiments using the same model recorders, mainly the Sony B100 and Sony B300. I purchased my recorders after hearing about his success with both of these recorders and with the belief that the recorder might make the difference in how EVPs are captured. With the recorders sitting side-by-side in various locations, I was only able to pick up a couple of voices where Rick would pick up many more.

It's quite possible that Rick's imprinted energy on the recorder may draw the spirits to his devices. One thing that Rick does differently than many other researcher is to pick a location and do a constant investigation. He believes that repeated attempts over time in the same location will yield better results by opening a rapport and making the spirits comfortable with his presence. So far, his efforts have yielded more EVPs from one location than I have captured during my entire career as a paranormal researcher.

Another major theory, which is the one I am putting my money on, is resonance. This basically deals with how sound travels in the form of vibrations. You are familiar with one form of resonance, which is the speaker found on your television or home stereo. The overall size of the speaker has a lot to do with its sound quality and how well it projects sound.

If you were to remove the speaker from its enclosure, the volume and quality of the sound will be completely different and sound very tinny, lacking depth and bass. What you have basically done is remove the speaker's ability to amplify and vibrate properly, which affected its sound quality.

This process also works in reverse when recording sound. The case surrounding the internal components of the recorder or microphone can help with the depth of sound being recorded, but can also allow sounds you cannot hear to get to the microphone. Just like the test we did earlier with rubbing your finger next to your ear canal, rubbing the case of the microphone will yield interesting noises during recording.

A spirit may not be strong enough to create an audible noise, but it may be able to resonate the case of the microphone or recorder enough for the internal microphone to pick up the vibrations. In this case, you would not hear the sounds until you play back the tape.

Tiger Electronics makes an interesting toothbrush for kids. It's a small digital audio player connected to the bristles of the brush. During brushing, the sound travels through the bristles and into your teeth, in turn resonating the sound through your head. I thought this may be a great way to test a recorder.

The toothbrush has a speaker inside, which uses pressure on the head of the toothbrush to make the volume louder the harder you brush. Without pressure the sound is barely audible. Without applying pressure to the brush, I held it against the bottom of my digital recorder on the opposite end to where the microphone is and talked normally into the microphone.

During playback you can hear my voice, but you can also hear the resonating toothbrush tune loud and clear almost overpowering what I am saying into the microphone. I also tried this test by putting the recorder at one end of my kitchen table and the toothbrush at the other. I was able to record the sound vibrations loud and clear, but I could barely hear them while they were being recorded.

The resonance theory seems to be the most logical and also explains how it would work with all recorders, past and present, including the mechanical recorder used by Waldemer Bogoras.

Until we are able to consistently and reliably record EVP, how they get into the recorder remains a mystery.

13. DIRECT RADIO VOICE

Communicating with the other side through the use of radio devices has been speculated about, attempted and, in some cases, believed to be successful. Using radio devices may seem like an odd idea, but it's not really as far fetched as it may sound. This is not to say spirits have radio transmitters on the other side, but spirits may be a form of a radio transmitter themselves. This idea partially stems from the law of conservation of energy and also from the study of the human brain.

The law of conservation of energy states that energy may neither be created nor destroyed but can change its form. To give you a simple example, if you turn on a flashlight, the stored battery power is converted to heat which, in turn, creates light. Although the battery may run out of power, the energy was not destroyed but instead changed to heat.

The human brain produces waves much like an AM radio transmitter but on a much lower frequency, around 3 to 12 Hz. It's speculated when the spirit leaves the body, it most likely retains the same energy

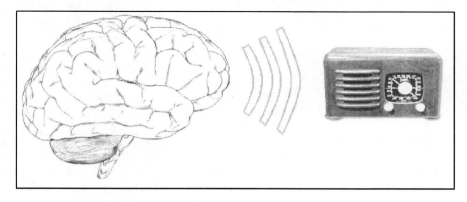

patterns. A disembodied spirit without the shielding of the human body may be able to affect a radio device that can receive these frequencies.

Although an AM radio is only designed to receive signals within a given frequency range, radios can also pick up on harmonic frequencies, which are weak offsets of the base frequency carrying the same information. If you have ever driven close to a radio tower, the station in question will bleed onto different parts of the radio band, allowing you to pick up the station at various spots on the dial besides its base frequency. Based on this phenomenon, a spirit resonating at 10Hz could possibly be picked up very faintly on a radio tuned to 550Hz.

Human brain waves are constantly changing based on our mood, thought patterns and activity. Spirit energy may also be constantly changing, unlike a radio broadcast, which is transmitted at constant frequency. Based on this theory, it would probably be impossible for a spirit to interfere with a radio that is tuned to a broadcast radio station, but it may be able to interact with a radio tuned to a blank station, as we just discussed. But since the spirit's energy is constantly changing, it may only be able to get through a word or two.

In the early '60s Latvian psychologist Dr. Konstantin Raudive started working with EVP after reading Friedrich Jürgenson's book and meeting him to learn his recording techniques. Raudive expanded his research by using an AM radio tuned to a blank station. This radio method creates a white noise background, which is said to help the spirit voices come through on the recording. This method promised better contact, but it also left many skeptics with the conviction that Raudive was just picking up stray radio signals from nearby broadcast stations.

Raudive worked with various electronics experts on different methods to attempt to make the voices more intelligible and clear. Instead of using the audio from an AM radio, he devised a technique using the heart of an AM radio, which is a germanium diode in place of a typical microphone. This is known as the Raudive Diode Method. Raudive claimed to receive the clearest communications this way. The skeptics of the time still claimed this device could pick up stray radio stations.

Raudive with radio apparatus

A germanium diode is capable of taking a tuned radio signal and converting it to an audible signal with the help of an amplifier and headphones. This simple device can be used to receive just about any kind of radio signal regardless of the band (AM, FM, LW, etc.).

Germanium Diode

If spirit energy retains the frequency range of the human brain once it leaves the body, it is possible that this device may also pick up on this energy and convert it to audible sounds.

Experimenting with this method is not too terribly difficult and any audio recorder with a microphone input will work without any soldering required. What you will need is the right jack for your recorder type, some alligator clips and a germanium diode. If this sounds hard, it's not. If you can pin a bag of chips closed with a clothes pin, you can build this simple apparatus to attempt communication using a diode.

First, you need to figure out what size jack you need for your recorder. There are two different sizes, 1/8 inch and 3/32 inch, the latter being more common these days. There is actually a third size that is rarely used except for high-end recording equipment, which is 1/4 inch. To figure it out, see if the tip of a shoelace will fit into the mic input of your recorder. You don't have to physically shove it in, but if it looks pretty close to fitting, then it's 1/8 inch. If the hole is too small, then it's 3/32 inch.

Head out to Radio Shack and ask for a bag of alligator clips with a 14-inch wire that has clips on both ends and a set of two conductor phono plugs in the size you figured out earlier.

Aligator Clips – Radio Shack stock number 278-1156.
1/8 phono plug – Radio Shack stock number 274-287
3/32 phono plug – Radio Shack stock number 274-290

You may ask if they have a 1N34A Germanium diode. Some stores may still carry these. If the store does not carry the diode, you can try another electronic parts store or get one online at various electronics places by searching for 1N34A. They should not cost more than $2 with shipping.

Putting it together is fairly simple. Unscrew the back end of the phono plug to expose the two metal conductors.

Spread the connectors so they are at a 45-degree angle from the length of the plug.

Take two alligator clip wires and connect one to each metal connector of the phono plug. Bend the diode leads so the whole thing is in a U shape then take the other two ends of the alligator clips and put one on each lead of the diode.

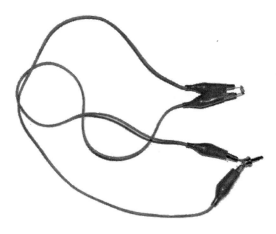

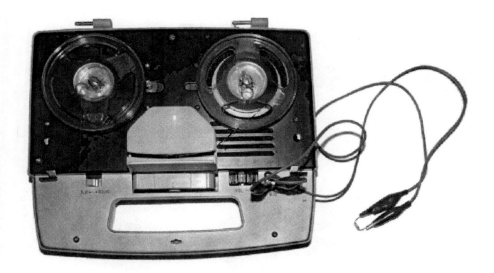

Plug this into your recorder and you are all set. The type of recorder is not crucial; just about anything with a microphone jack will work.

It is best to leave the recorder in one location while using this method. The connections can be rather noisy if the diode gets moved around and may cause a false positive. Also, be sure the diode is not lying on any conductive surface like a metal table; this may ground the diode, which will yield no results on playback.

During playback of your recording session, you will not hear yourself talking or asking questions, but you may be able to hear spirit voices. This makes this method nice as you know there are no stray sounds that can interact with the recorder. The obvious problem is that you can't hear yourself asking questions, so it is best to write them down to remember. If you are working with this method anywhere near a radio tower, you may pick up stray radio signals. That's why you will need to be aware of your surroundings.

Raudive used several different variations of the diode method incorporating a few extra parts here and there and even attempting to attenuate the signal for use with spirit communication, but the previous method appeared to be the most fruitful.

14. PSYCHOFON

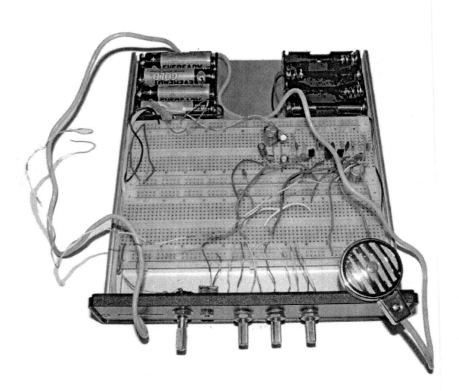

In the early '70s, Viennese engineer Franz Seidl created a new apparatus which would expand the diode technique to another level. Seidl called his invention the Psychofon or sometimes the Psychophone. This device incorporated several amplification circuits around the germanium diode and a microphone, allowing the researcher to also hear his own voice on the recording, unlike the original diode method.

Seidl claimed great success with this device, recording thousands of voices, which he outlines in his 1982 book, "Registration of Supernatural

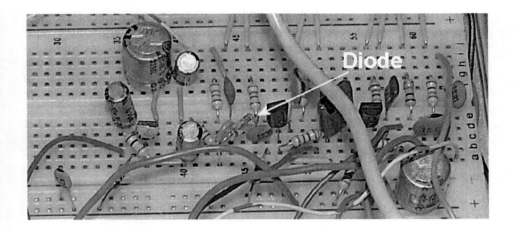

Voices." This book included his experiences and documentation for constructing the Psychofon. Raudive also used this device during his experiments, claiming great success as well.

This apparatus is basically a detuned broadband radio receiver just like Raudive's Diode but it incorporates amplification to help the spirits come through. Unfortunately, it also picks up various AM radio stations, making it difficult to discern spirit voices from radio personalities. It's kind of like listening to chatter in a school cafeteria during lunch.

Constructing this device is a bit different than just buying a few parts at Radio Shack and clipping them together. A hobbyist background in electronics would be helpful, but anyone can build this device pretty easily with a little time and effort. Radio Shack sells a hobby kit that will give you just about everything you need to build it and experiment with it without having to solder anything.

The hobby board is Radio Shack stock number 28-280. It

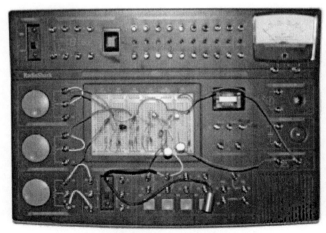

Radio Shack Hobby Board

comes with a bunch of parts and connector wires. If you decide to give this a try and have never worked with electronics before, you may want to try a few of the sample projects given with the device to get a feel for how this works and an idea of how to read schematics. You can find the schematics for this device in the back of this book along with the list of parts needed. You can also find other various schematics for this device available on the Internet.

    Any skeptic is going to claim the voices being received are just a conglomeration of the radio stations, creating a matrixing effect that sounds like possible spirit communication.

15. SPIRICOM

Finding the right frequency for direct spirit communication was on the minds of several researchers throughout the '70s and early '80s. As audio recorders rolled in various countries, exploring the realm of the unknown, George Meek and his colleague William O'Neil were hard at work on a new device, hoping to give a fresh perspective on direct two-way communication.

Meek had a different idea of how the spirit world is organized. His theory was based on six different levels starting with planet Earth and working its way up through the lowest astral, middle astral, upper astral, causal and mental and, finally, celestial. Spirits supposedly spend different amounts of time in each of these levels as they ascend to the top. Each level has a distinct frequency, which Meek believed he could reach through a radio device and engage in two-way communication with the inhabitants.

Meek worked on several systems in the early '70s that yielded minimal results. During the early days, he consulted with various psychic mediums claiming contact with spirits who gave advice and guidance on how to modify and adjust his devices for spirit communication, but none of the spirits were able to actually speak through the devices.

In 1975, Meek started work on a project with psychic medium William O'Neil, who had a background in radio communication. The system would originally be known as the Spiricom Mark III (it became the Mark IV in the later years.) As work on the Mark III/IV device got underway, O'Neil came into contact with several spirits, which would be instrumental in successfully conversing with the other side.

Unlike the diode method and the Psychophone, which are just broadband radio receivers, the Spiricom was a full communication system that offered a means for vocal conversation. The new system would eventually yield some 20 hours of direct communication with a spirit claiming to be the late Dr. George Jeffries Mueller who reportedly entered the ethereal world on May 31, 1967.

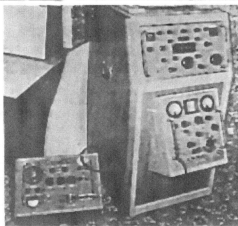

Spiricom Mark III

The first step in the process was an electronic tone generator creating 13 separate frequencies. The spirit that identified itself as Dr. Mueller helped pick the frequencies that together make up the vocal range of the typical male voice.

The tone generator connected to a low power amateur radio transmitter with a short quarter wave antenna tuned to 29.570 Mhz on the AM band. The transmission distance was very short and confined to the laboratory.

The receiver was tuned to the same frequency with a similar quarter wave antenna for short-range reception. The antennas were placed about six feet apart. An external speaker connected to the receiver was placed on one side of the room with a microphone and tape recorder on the other. The entire room was used as an echo chamber.

It is speculated that the transmitted tones were able to break through to the ethereal plane, allowing a spirit to interact with the radio waves and tones resonating its speech onto the signal. The modified signal is received and speech is heard through the external speaker. Basically, it's a way of providing vocal chords for the spirit world.

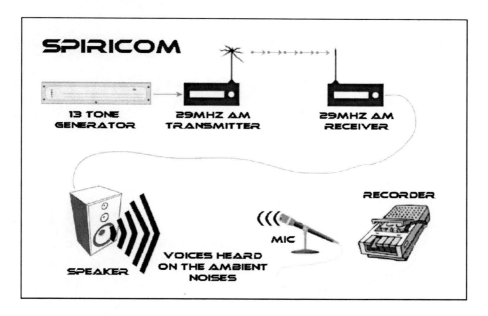

After many attempts and adjustments, the spirit calling itself Dr. Mueller was able to talk through the device with the help of William O'Neil on April 16, 1980. Mueller's voice was very robotic. O'Neil recorded many hours of conversation with the so-called Dr. Mueller. These communications ended abruptly after a few short months. Meek speculated that Dr. Muller had advanced to the next level, which was beyond the frequency range of the Mark III/IV system.

Though Meek and other colleagues attempted communication using the Mark III/IV, Dr. Muller was only able to make contact in the presence of O'Neil. Mediumistic abilities appeared to play a big role in the proper operation of the device.

Meek published a manual in February of 1982 entitled "Spiricom: An

Electromagnetic-Etheric Systems Approach to Communications with Other Levels of Human Consciousness." This was originally accompanied by an audio cassette explaining his approach and examples of real two-way communication. Hard copies are available without the cassette if you do some searching. You can also find audio files and the complete manual at www.worlditc.org and various other places on the Internet.

16. GHOST BOXES

Having the right combination of diodes, resistors, integrated circuits, wire and various other electronic components may not be the only thing needed to break through to the other side. It may also require a level of respect for your receiving party and a little bit of psychic amplification on your own part.

In mid-2002 some interesting documentation hit the Internet with information on how to construct a device that would become known as "Frank's Box," named after its creator, Frank Sumption. Much like the Spiricom and Psychofon, Frank's Box uses radio technology to help open communication lines to other planes, but it appears to be having greater success than the earlier devices.

Sumption began experimenting with EVP around the year 2001, which led to using a computer program called EVP Maker, invented by German researcher Stefan Bion. After receiving various messages from "computer savvy" spirits relaying messages for other spirits who were not so technologically advanced, Sumption came up with the idea to create a device that all spirits could use. His design was apparent to him almost immediately, but the actual construction of the device has led him to create at least 27 different models at the time of writing this book. Each box is unique in design and construction, but all are based on the same principle.

Sumption's spirit receiver starts off with a white noise generator that is fed through a random voltage circuit of his own design. The random voltage circuit is linked to an AM or FM radio receiver from a late '80s/early '90s car stereo, which reacts to the voltage by tuning to a specific spot on the radio dial. This is known as voltage tuning and is a common function of these car radio receivers. The received radio station is amplified and piped to a small sound chamber in the back of the box, which creates an area for a spirit to interact with the audio. A microphone pickup in the chamber is amplified and sent out through a monitor speaker on the front of the box.

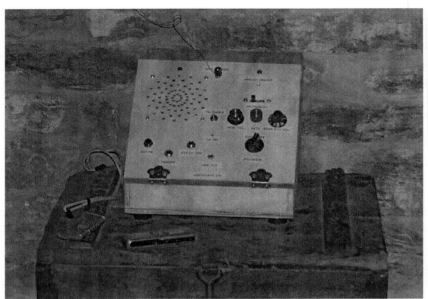

Frank's Ghost Box

Though various radio stations are turned in for a split second every so often along with regular static, the device also allows the spirits to interact with the device in various ways to create their own vocals through the receiver and for lack of a better term, talk through the device.

A newer version of the box simply scans back and fourth through the AM or FM band which Sumption calls the "sweep" method. At first, he believed the random voltage design was what allowed the device to work but after using the sweep method, he has since changed his mind as it seems to do a better job. Sumption has made his plans available on the Internet for anyone who is interested in experimenting with his device. For those who want to take it a step further and create the entire box from start to finish, he also has his own radio receiver plans, which replace the need for an old car stereo.

Sumption has created at least 27 versions of the box to date and given them to several individuals for ongoing tests. The initial results have been pretty positive and many people have experienced some kind of communication that they regard as evidence the box really works. Unfortunately, the difficult part about Sumption's design is his use of radio bands as the medium for receiving the voices, which could lead any skeptic into debunking the operation of the device.

Frank's Box will receive little snippets from various radio stations as it scans through the AM or FM band. At any given moment the device may

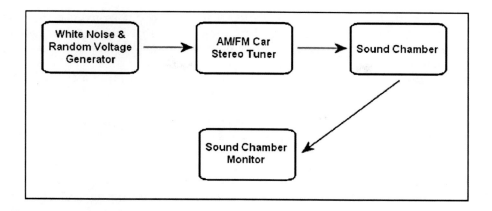

spew some words from passing stations, which could be put together as a message and claimed to be from a spirit. From a skeptic's point of view, it would most likely be looked upon as audio matrixing.

Another possible scenario includes a few parts from Radio Shack and a couple of minutes of assembly, which could yield a small yet powerful enough transmitter to broadcast over the AM or FM band and inject various words and phrases into the box directly. Definite care needs be taken when operating the device to ensure the above scenarios are not part of the equation. Using recorders and other tools, such as an EMF detector, can help legitimize the results. EMF detectors should be placed far enough away as not to cause interference with the box or produce false readings on the detector itself.

Sumption does not condone the use of his box for paranormal investigations, though some peoplE have been attempting communication at various famous haunted locations with a bit of success. Sumption feels taking the box into a haunted location either biases the evidence or may induce expectations of talking to a specific spirit. He also states that using the box in locations where radio signals are sparse, may hinder the performance of the box.

If you are not into designing and creating your own circuits, there is a way to try this technique without actually building anything, though it may not be as effective and the voices may not be as clear. The technique is called the manual sweep method and only requires a radio with a tuning knob and some kind of audio recorder.

To use the manual tuning method, get yourself comfortable and clear your mind. Don't expect to hear anything specific from any particular spirit. Start recording and announce you are doing an EVP session and invite any spirits who wish to communicate to do so through the radio. Tune the radio back and fourth across the AM or FM band limiting the sweep length from about three to five seconds to get from one side of the band to the other. Continually sweep through the band for about five minutes. You most likely will not hear the voices in real time though you may pick up on a word or two.

When playing back the recording, you may hear actual voices that are not the result of bypassing radio stations or you may hear a conglomeration of stations that have been put together to make a short sentence. Listen carefully as some messages may not be intelligible to start but may make more sense if you play the recording back a few times.

Sumption administers a message board on Yahoo called EVP-ITC for those wishing to share their experiences and audio clips. He also posts his new designs with information of what he is working on and various construction methods. Hearing the box speak on this site is quite interesting and can give you an idea of what you might hear by attempting the manual sweep method.

If you decide to go searching the Internet for Frank's Ghost Box, you may find several individuals making some pretty substantial claims as to who they have contacted or who can and cannot use the box. Many of these claims have no substantial evidence to back them up and may not be endorsed by Frank Sumption. He has maintained that anyone should be able to use the device although some may have better results than others. Properly operating the device may be directly linked to the users' psychic awareness and sensitivity to spirits. Sumption has stated that his box is not intended to be a direct link to specific spirits or to be used as a mediumistic device.

Various clone boxes have emerged on the scene claiming similar results to what Sumption has received. Another version is called Joe's Box, which is based on the same concept as Frank's Box, but uses much different components. Joe's Box does not incorporate the use of a sound chamber, which makes the device a bit smaller than the typical Frank's Box. I, myself, have also created a smaller version of Frank's Box, which is based on his plans using an old car stereo tuner with a few different modifications and parts in the circuitry than his original. I have also created a circuit to modify an existing boombox-style radio. Plans are available on my website (www.ghost-tech.com).

Radio Shack sells a small portable radio for about $24 that can be modified to work much like a Frank's Box. This modification does not take any technical knowledge of radio electronics or any ability to solder. If you can use a screwdriver, you can do this modification. This also does not change the radio's ability to work in the normal fashion.

Pick up a 12-469 radio from your local Radio Shack. These are normally hanging on the wall near the headphones. It's a basic AM/FM radio with headphones. You will also need a small Phillips screwdriver

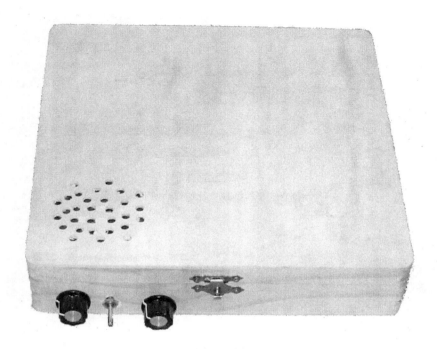

Craigs Box (Franks Box Wannabe)

and a small pair of pliers.

Remove the screws from the back. There are two visible on the back and one in the battery compartment.

Carefully pry the back apart. There are a couple of snaps around the outer edge that should give way pretty easily, but do not pull too hard; there are wires connected to the back that might break from their solder points. Remove the two screws from the inside PC board.

Carefully separate the top PC board from the bottom. Be gentle; you may need to pull the plastic top, where the headphone plugs in, out just a bit.

On the board you just removed from the radio, you will see a row of pins

Radio Shack 12-469

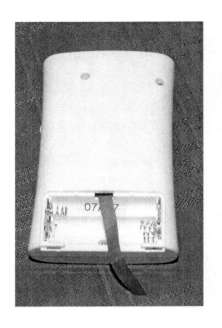
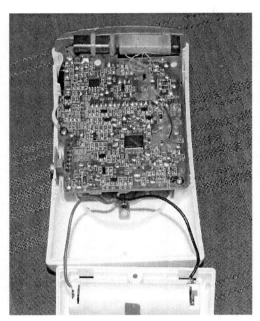

that have small labels on them. Find the pin with the word "mute" right under it. You can either bend this pin down and out of the way like I did, or just remove it all together.

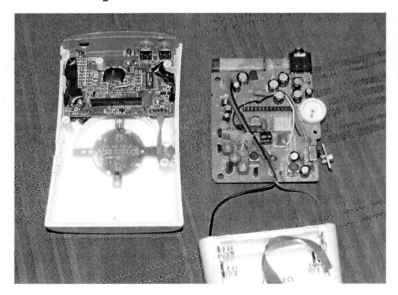

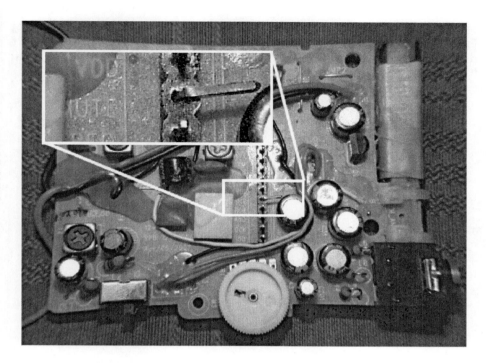

Put the board back into the radio. This can be a little tricky. Be sure the pins line up by looking between the boards. Once they are lined up, carefully push them back together. Put the screws back into the board. Snap the back on and put the screws in. Now you are ready to run your Franks Box clone. Put in some fresh AAA batteries and fire it up!

Push and hold the up or down button on the radio. This will cause the unit to start scanning through the radio stations. With the mute capabilities removed, the radio will continually scan without stopping. Turn on your tape recorder and start listening for voices.

If you are not looking to modify or build a device, a company called Paranormal Systems (www.paranormalsystems.com) is working on a production version that should be available sometime in early 2008 called the Minibox.

The Minibox works much like Frank

The Minibox

Sumption's original design and has been given rave reviews by many Frank's Box enthusiasts including Frank himself. As with any device designed to communicate with spirits, time and patience are your biggest assets. You cannot expect instant results.

17. GHOST DETECTION

Finding a ghost is not an easy endeavor and typically requires patience and something which is sensitive to its presence. It's speculated spirits give off energy that can interfere with or be detected by radios, televisions and other electromagnetic receiving devices. The main problem with a standard receiver like a radio is that it's designed to only pick up a very small portion of the electromagnetic spectrum through the tuning mechanism and reject all other frequencies.

Imagine standing in an abandoned sanitarium trying to tune in spirit voices using an AM radio for hours and getting nothing. Kind of frustrating and time consuming isn't it?

Now imagine a spirit standing three inches from the radio screaming at it and passing over and through it for hours trying to make contact with no results. Makes you wonder who might be more frustrated.

In the early 1920s the American Psychic Institute and Laboratory developed a device to detect the presence of disembodied spirits called the Uluometer. Using a large coil of copper wire and some other basic circuitry, the device was able to pick up the electromagnetic energies emitted by humans and supposedly those emitted by spirits. To announce the presence of this energy, the device

The Uluometer

SNUFF OUT EGG CANDLING

GARY, Ind. (INS) The national patent council says eggs are tested today by electronics for internal quality and potential perishability under a new invention. The egg is placed in a coil lying in an electromagnetic field of radio frequency. Good eggs absorb the least power, bad eggs the most. Poultrymen consider the method close to 100 per cent accurate. They say the invention will mean a large saving over present system of candling egges.

would make a howling noise which changed in pitch the closer the energy was to the copper coil. This noise would earn the Uluometer the nickname "psychic howler."

The Uluometer was most likely the first electromagnetic apparatus designed for ghost investigations. Similar devices of the time were in use for other forms of detection such as finding precious metals by emitting and detecting changes in electromagnetic waves.

Electromagnetic detectors eventually found their place in different industries around the country including medicine, mining and even on the farm to check the potential perishability of eggs. At the time, very little was known about electromagnetic fields, which were of little concern to the common person.

In the early 1970s new studies were being conducted to find a link between electromagnetic fields and various health problems. Power lines were the biggest concern of the time, but just about any electronic device can produce some kind of electromagnetic field that may be considered dangerous.

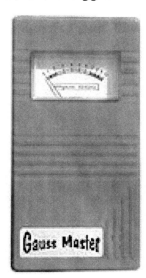

By the late 1990s the health industry was going crazy over the potential side effects of electromagnetic radiation. Many electronics manufactures jumped on the band wagon and started offering portable Electromagnetic Field Detectors (EMF Detectors). Most of these devices are designed to pick up normal house power and other forms of low frequency energy and give a metered readout of the possible dangers imposed. Higher priced units can also pick up cell phones and other high frequency devices.

By the late 1990s, EMF detectors were in use all over the world and also found a new home in ghost hunting kits. Much like the Uluometer, EMF detectors may be capable of picking up on the energy emitted by spirits.

EMF detectors are really not much different than the Psychophone and are basically broad band receivers. The big difference is that EMF detectors are designed to give a strength reading of how great the energy is and not an audible representation of the energy itself.

EMF is measured in a unit of magnetism called Gauss which was theorized and discovered in the early 1800s by Johann Carl Friedrich Gauss. Using his methods of calculation, magnetic pull could be represented in terms of mass, length and time.

Most detectors have a standard frequency range of 30 Hertz to 300 Hertz, which is considered Extremely Low Frequencies (ELF). Most major appliances, televisions and other electronic devices emit some kind of EMF in the ELF range. The broad frequency range poses a problem with using these detectors for paranormal investigations since there is no way to tell whether you are picking up AC house power, a natural EMF or a spirit.

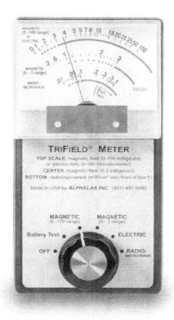

Tri-Axis EMF Detector

There are many different portable detectors available on the market but there are really only two categories: single axis and tri-axis. Most meters are only single axis and require three measurements at different angles to get an accurate reading. Tri-axis meters only require one reading and give instant results when positioned correctly.

To take a reading with a tri-axis meter, the device needs to be positioned in such a way as to get the highest possible reading on the device. To do this, start with the device straight up and down and walk around until you get a reading, then roll the device in different directions to find the highest possible reading. This is the field's strength.

To get an reading with a single axis device, use the same principle as the tri-axis and find the highest possible reading and write it down, then turn the device 90 degrees to the left to get the next reading then 90 degrees up to get the third reading. Once you have all three readings, you will need to do an equation to get the actual strength.

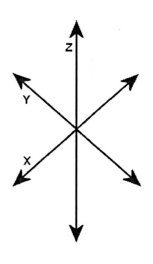

Strength = $\sqrt{(X^2 + Y^2 + Z^2)}$

Paranormal investigators are looking for a spiking inconsistent reading on their device as apposed to keeping track of the actual field strength, so using the single axis meter for investigations is normally sufficient. The main problem with a single axis meter is you may pass over a field and not recognize it because the meter is not at the right angle or pointed in the correct direction.

For a simple test, I used my KII meter, which is currently a hot item on the market and pretty simple to use. This is a single axis meter with five lights that depict the danger of the electromagnetic field. Believe it or not, a standard clock radio gives off a good bit of EMF at a short distance, which makes it a perfect subject.

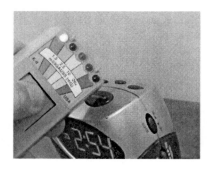

With the meter hovering just above the clock radio, I get full tilt on the KII display, which is telling me the device is putting out a dangerous level of EMF if I am about an inch from it.

Interestingly enough, if I tilt the device 90 degrees at the same spot on the clock radio, the EMF field disappears. The field didn't really go anywhere; I just changed the angle of the sensor inside the device until it was is no longer in flux with the electromagnetic field of the clock radio. Definite care should be used while investigating with an EMF detector. A tri-axis meter will not have this problem but the strength will vary depending on its angle.

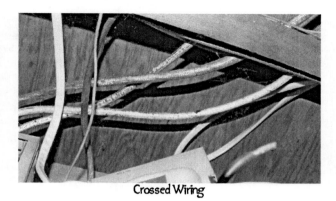
Crossed Wiring

Many home appliances like refrigerators and air conditioners can cause an EMF field through the house wiring while they are running that will disappear when the appliance stops running. Crossed wiring in the basement, under the floor and in the walls can cause an EMF field, which will travel the entire length of the wire. Depending on how much power the connected device requires to operate will define the level of EMF emitted by the wire.

To effectively use an EMF detector during a paranormal investigation, you will want to take a base reading at different angles throughout the area being investigated and look for potential hot spots and their root cause. Be sure to take into account air conditioners, heating units and any other device that can switch on and off by itself.

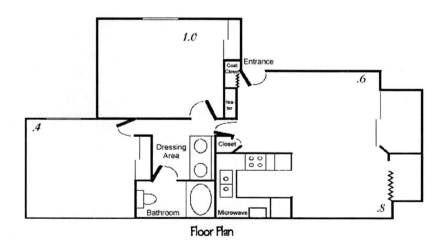
Floor Plan

A floor plan of the area can help, but if one is not available a simple sketch or at least a description of where the EM fields are located can help later on when the investigation is in full swing. Just remember to rotate the device at different angles when you find a hot spot so that you get the highest possible reading.

During an investigation, a significant change in the EMF reading can indicate possible activity, but nothing is foolproof. The addition of a frequency counter can help solidify the evidence. A frequency counter can tell the difference between AC voltages running through the location or a possible anomaly on another frequency.

A basic frequency counter

This device works much like an EMF detector except it will give you the frequency reading of the field instead of the strength. Anything related to general house power at 120 volts AC will give you a reading around 60hz. This includes appliances like a refrigerator, television, air conditioner, etc.

A useable frequency counter should have a range of 1hz to 300hz or better. If you receive a spike on the EMF detector and the frequency counter is reading about 60hz, you are picking up regular 120 volt AC power. Lower frequencies between 1 to 10hz may be a good indication of spirit activity.

Keep in mind that these devices are not designed for paranormal investigations and it will take time and practice to use them effectively.

18 OTHER POPULAR DEVICES

Temperature Probe

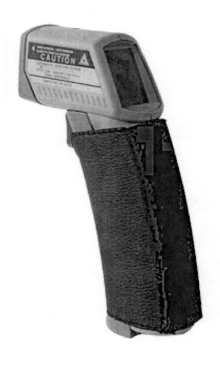

Not knowing where spirits fall on the electromagnetic spectrum opens the door for various sensing and detecting devices to be used during an investigation. If it senses some kind of change or electrical disturbance, it's probably been used to try and detect paranormal activity.

Probably the most over-used yet misunderstood device is the non-contact temperature probe. It is speculated that when a spirit is attempting to manifest, it will draw energy from its surroundings, which can lower the ambient air temperature around it. Sudden temperature drops can be a good indication that you are not alone.

A non-contact temperature probe uses infrared radiation emitted by an object to gather its overall temperature. With the push of a button, the device will display the temperature on a small LCD display. The aid of a laser pointer built into the front of the device makes pinpointing an area to read quite simple.

The problem with these probes is their accuracy, which severely degrades within a few inches from the device. The farther away you are from the object in question, the greater the area is scanned by the probe. Though they may seem like a good idea, these devices were not

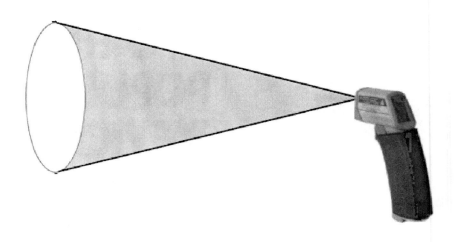

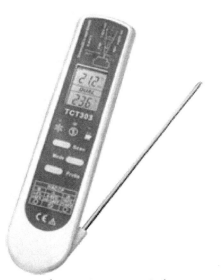

Ambient Air Temperature Probe

designed to take ambient air temperature readings, which is typically what we are interested in during an investigation.

If you are considering using one of these devices during your investigations, buy one with an ambient air temperature probe. This will give you a far more accurate reading and make for better evidence gathering. Keep in mind that heat rises. When using this device, sudden changes in temperature can be due to where you are pointing it. Keep it at a constant height and move the device slowly to avoid breezes over the probe, which will give you an inaccurate reading.

Ion Detector

Another common theory about spirit manifestation is the negative ion count in the air will increase or the positive ion count will decrease. Various counters are available on the market that can give both the positive and negative ion count through a digital readout. Just like an

EMF detector, a base reading can be taken before the start of an investigation and then changes recorded throughout the investigation.

Geiger Counter

Although radiation is not a concern during a paranormal investigation, that is unless you are checking out the Three Mile Island Nuclear Power Plant, some folks have attempted to use a Geiger counter as a means to detect possible spirit activity.

Digital Ion Counter

These can be found on various auction sites through the Internet and in military surplus stores. Geiger counters are designed to detect the decay of nuclear particles and various other forms of harmful radiation. This is basically an EMF detector for a much higher frequency. A simple needle or digital readout displays the possible threat of contamination.

Geiger Counter

Unless spirits emit energy around the same frequency as X-rays and gamma rays on the electromagnetic scale, a Geiger counter will be useless during a paranormal investigation, but I guess it never hurts to give it a shot if you have one sitting around in your basement collecting dust.

Motion Sensor

Using a motion detector during your investigation is another possible way to capture evidence. A typical detector senses a sudden change in temperature in front of the device by monitoring infrared levels. When a significant change is detected, the device notifies the main alarm panel,

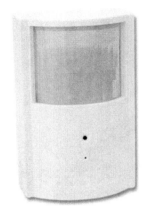

Motion Detector

Game Camera

which sets off the alarm.

Personal detectors are available on the consumer market that are standalone devices with their own internal alarm system. These can be set up in an unoccupied room with a video camera to capture any activity that may have set off the alarm.

Motion sensors are also being used in conjunction with digital still cameras to take a snap shot of the detected presence. These are typically known as game cameras and can be mounted to a tree and used to capture images of deer and other animals passing by.

Most game cameras are designed to operate for weeks at a time and an excellent choice for ongoing investigations when you can leave the device in a secure unoccupied location.

Radar Detector

Another odd device which does not get to much attention is the radar detector. As strange as it may sound, a normal car radar detector may sense the presence of paranormal activity.

This device is designed to receive radio signals emitted by a police radar gun. When an officer alongside the road pushes the button on his radar gun, it emits frequencies that are at the higher end of the microwave band. These signals travel out to your car and bounce back to the officer's gun, which calculates the speed at which you are traveling.

Although it's designed to save your tail when you are flying down the highway, a radar detector is basically just a radio receiver for the microwave band, which ma be able to pick up frequencies emitted by spirits.

Compass

Probably one of the oldest methods in the book is to use a compass during your investigations. It is speculated that spirits can disrupt the magnetic properties of these devices and cause the plate to spin or point in another direction.

The theory that spirits are magnetic has been around for probably as long as compasses themselves, but recent technology is starting to prove that this is most likely not the case. Although a compass is based on magnetic pull, it may be that the spirit is able to manipulate the dial based on psychic energy instead of magnetic energy.

Place the compass out in the open where no foreign objects can disrupt its normal operation. Once the dial settles to pointing north, ask the spirit(s) to interact with the device and try to move the dial or spin it completely around, if possible. As simple as this may sound, video footage of a spinning compass can make great evidence.

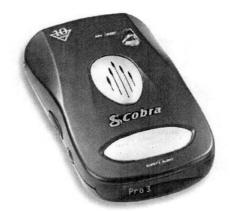

Radar Detector

Compass

19. BUILDABLE PARANORMAL GADGETS

Building your own gadgets is not as difficult as it may seem. If you invest a little time and effort into the construction, the outcome can be rewarding and also give your investigations a unique aspect. Unfortunately I cannot give you a course on how to build electronics, but much information can be found online and from your local library, which can teach you the basics, no Ph.D. required.

As I mentioned earlier, Radio Shack sells a nice little electronics hobby board, which can be used for many different experiments and is portable enough to be taken along when you want to try something out before you solder it together. This kit also comes with basic projects to teach you how to read schematics and how to properly wire the projects. The only thing it does not really teach is how to solder, which is also not as difficult as some would lead you to believe. Information and videos are available on the Internet which can teach you how to solder properly and safely.

The following schematics are rather simple and only require a few parts but can be a great asset on an investigation. These and other schematics are available on my website at www.ghost-tech.com under the schematics section. You may find other similar projects available on the Internet by searching for "paranormal schematics." Parts, project cases and various other information can be found at www.bgmicro.com, www.mouser.com cand www.allelectronics.com as well as local and online electronics distributors.

Negative Ion Detector

Detecting negative ions is a newer technique for paranormal investigations and one that is questionable. It is speculated that the amount of ions in an area will rise in the presence of spirit activity. Here is an incredibly simple project that will allow you to experiment with negative ions without spending a ton of money on a store-bought

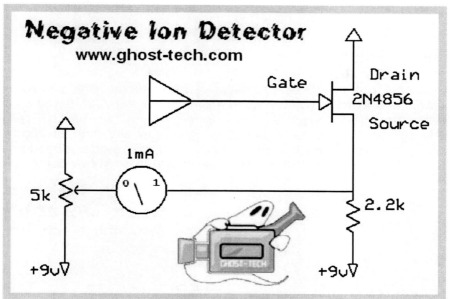

Negative Ion Detector Schematic

detector.

This device is based on a 2N4856 field effect transistor. Unfortunately, you will not find one of these in the Radio Shack hobby kit, but they can be found online for a dollar or two. The antenna can be an unshielded piece of wire about three to four inches long.

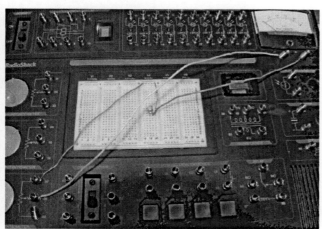

Negative ion detector on the hobby board

Testing the device is quite simple, if you a have an Ionic Breeze or similar air cleaning device, this will set off your detector. Another simple test is to create static electricity by combing your hair or petting an animal near the device. The greater the concentration of ions, the higher the meter reading.

Simple EMF Detector

Using the same parts as the negative ion detector, plus adding two resistors and a diode, you can build a simple EMF detector. Based on the 2N4856 field effect transistor, this device can pick up on EMF spikes pretty reliably, but don't expect it to give you any kind of accurate reading. Be sure the diode is connected in the proper direction as discussed in the documentation that comes with your hobby board.

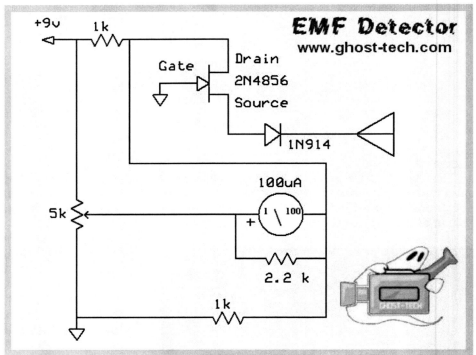

Simple EMF Detector

Test the device in front of your television. When the TV is off, the device should not register any EMF, unless the TV was just turned off, then it may show a little EMF. Turn on your television and the device should peg the needle then settle down to a constant level.

Temperature Change Detector

Now for something a little more complicated. It is speculated that spirits will rapidly change the temperature of a given area when passing through or attempting to manifest. The following device can detect slight changes in ambient air temperature.

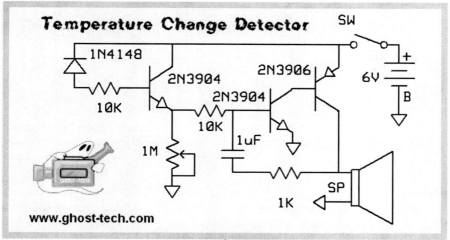

Temperature Change Detector

You should have all the parts you need with the Radio Shack hobby board. The diode is actually used as the temperature sensor. You should be able to substitute the 1N4148 diode for the 1N914 used in the last project.

You can use a hairdryer to test the device. Power the device and adjust the potentiometer so the device is quiet. Turn on the hairdryer and point it towards the diode from about a foot or two away. As the diode changes temperature, the device will start making noise, the greater the deviation in temperature, the higher in pitch the noise will become.

The Uluometer

Though no plans were ever released on how to make the Uluometer, this is my rendition of what I believe the device to be. You should have all of the parts needed to build this device except the TL082 OpAmp and the ferrite core coil. You should be able to use the OpAmps which came with your hobby board but you will need to find a coil. Both the correct OpAmps and the coil can be purchased at Radio Shack for a couple of

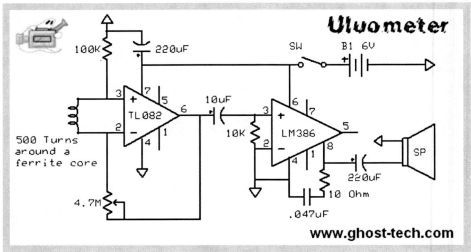

The Uluometer

dollars.

To test the Uluometer, move the coil near any device that is giving off EMF and you should hear sound coming from the speaker.

This device can be used side-by-side with an EMF detector to give a strength and audible representation of the EMF field in question. With a little practice, you can figure out which EMF fields are electrical and which are natural by listening to the sound through the speaker. This device may also be used in conjunction with an audio recorder for EVP research.

I found myself using this device so much that I decided to make a small version of it that I could carry along with me. The black compartment on the bottom holds four AA batteries. The coil at the top is removable so I can

Uluometer

put in longer coils or other parts like a germanium diode for testing. This one little device has a myriad of uses.

Frank's Box Linear Scan

As you advance in your skills, you may consider attempting to build a Frank's Box from the plans available online. Depending on which version you build, it may take a bit of time and troubleshooting to get it right. After building a few versions myself where I removed the radio component of a car stereo, I decided to try the reverse and put the linear circuit inside the car stereo instead.

I took a trip to the junkyard and pulled a few different older car stereos until I found that my favorite one to work with, due to its size, is the Chrysler AM/FM standard radio from a 1989 Caravan. This stereo was used in many years of Chrysler cars, trucks and vans and can be found online or in a junkyard for typically less than ten dollars.

You can pretty much use any car stereo from the '80s and early '90s, but you need to get one that is voltage tuning. Many of the newer car stereos use microprocessors to tune the stations, which will not work with the linear scan circuit. The easiest way to tell them apart is by the display. If the

Chrysler AM/FM Car radio

radio only displays the radio station in digital format and maybe a clock, it should be voltage tuning. If the display can print words or pictures, it's most likely digital tuning. Most foreign cars used voltage tuning in the '80s and early '90s. You can't go wrong with a Nissan or a Toyota radio.

Start off by building the linear scan circuit. You may want to build the circuit on your hobby board and get it working before soldering it together. You can test its function by using a multi-meter in a slow scan mode, which will make the voltage rise and fall at a slow pace.

Try to use the values of parts listed, any slight deviation may stop the circuit from scanning all the way through the radio band. Nine volts is sufficient, but for the car radio we will use 12 volts, which is what the stereo requires to operate. Don't worry about your circuit, we can

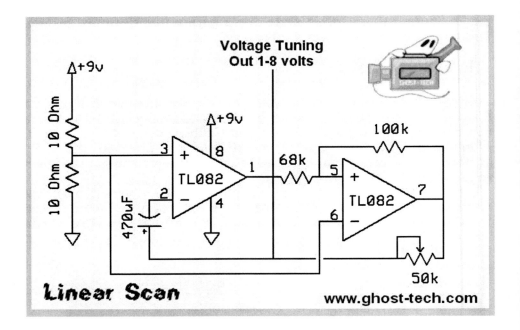

actually dump about 18 volts into this circuit before it can cause any damage.

Other parts you will need include a small eight ohm speaker. You can find these inside old radios or you can buy one online for a couple of dollars. The size is not important, unless you want it to fit inside the radio with the circuit.

Get a battery holder, which can hold eight AA batteries to generate the 12 volts needed for the stereo and the linear scan circuit. Again, if you want it to fit inside the car stereo, be aware of the size and shape. You will also need something that will suffice as an antenna. The length is not too terribly important, but get something that is at least four inches long or longer.

Get your circuit working on the hobby board then solder it all together on a proto board. Power it up and test again. Make sure it's working 100 percent.

Next, we will want to prep the radio for the circuit. This can be difficult or easy depending on the type of stereo you have. It's probably a good idea to test the car radio to make sure it's working. The eight AA batteries should be sufficient to drive the car stereo for a good bit of time. Hook up the speaker and the batteries and give it a try. If it's working, then it's time to take it apart and do some modifications.

Remove the panels of the radio to expose the bottom and the top. You

will want to locate the AM and the FM receiver modules. Sometimes these are in one long metal box and other times they are separate.

The Chrysler stereo has separate modules. The AM module is typically smaller than the FM module. Flip the radio so you can see the circuit board. If you are lucky, the pins will be labeled. If not, you will need to hunt for the correct pin.

Unfortunately, the Chrysler stereo is not labeled. To find the pin, you will need a multi-meter, preferably one with a needle instead of a digital readout. Most digital meters are slow to respond to changing voltages.

Hook the battery pack up to the radio and turn it on. Select AM radio to start and hit the scan feature so the radio is looking for stations. Select 0-10 DC Volts on your multi-meter. Hold the negative lead (black) of the multi-meter on the chassis someplace to ground it. Take the positive lead and hold it on each pin of the AM module for a couple of seconds. If the needle starts rising, then falls to the bottom then starts rising again, congratulations are in order; you have just found the voltage-tuning pin! The Chrysler pin is the third one when counting from the side of the faceplate across.

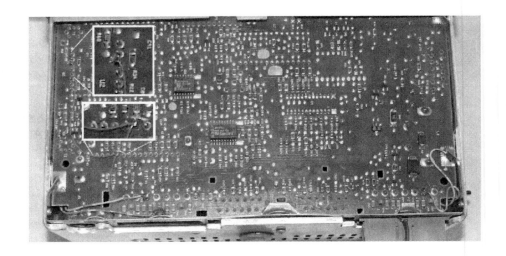

Do the same for the FM module.

Once you have found your pins, unhook the battery. Locate the circuit trace, which is leading to the pin. Take a knife and sever the connection so the radio no longer has control of the voltage tuning. Be careful not to sever any other connections. Once this is complete, hook the battery pack up and check that there is no voltage for the pins you just severed, if you find voltage, you did not sever the connection. Try again until there is no voltage.

Solder a wire between the two pins and another wire from one of the pins, which can be put through a hole in the pc board to go to the voltage tuning output of our linear circuit. You may also want to locate the power output and speaker connections and solder wires from those to the other side.

A hot glue gun can be your best friend when doing this kind of modification. I hot glued the speaker on the right side in the back, then ran the wires through and connected them to one of the speaker pins. I drilled a hole on the left side and put the 50k potentiometer through so I can adjust the speed. The main circuit is hot glued down front. Make sure you are not heating up any components while gluing.

If you plan on putting the batteries inside, make sure you have a way to disconnect them or add a switch of some sort so you can turn the whole thing off. Hook your battery pack up and test to be sure the circuit and radio are functioning before reassembling.

If you run into problems, use your multi-meter to be sure voltage is reaching your circuit and that the circuit is working.

Other modifications can be found on my webpage at www.ghost-tech.com along with more schematics and circuits. Remember to take your time and do it right the first time, De-soldering is not fun and typically makes a big mess. If you create the circuits on the hobby board

Chrysler Radio Franks Box Clone

first then transfer them over to a proto board, you will have much greater success.

20. WORKING WITH A SENSITIVE

Electronic devices are great on an investigation and may provide evidence of a haunting and photographs to go with it, but trying to get the whole story of why the spirits are there and who they may have been when they were alive is a much more difficult challenge. Though EVP may start to answer these questions, it's probably not going to give you everything, so it may be time to call in someone who can.

Incorporating the impressions of a sensitive into your investigations can help solidify possible evidence and may even help point out the hot spots where evidence can be obtained. It's kind of like having your own personal tour guide walking you through the location and explaining the history and possible feelings of the spirits.

Unfortunately, there is a bit of controversy involved with using sensitives during an investigation and some folks don't even believe the living can feel or communicate with spirits. This goes hand-in-hand with believing or disbelieving in spirits altogether. Some people think if you believe in ghosts, you have to believe in psychics as well. This is something that could be debated for years without resolution and is basically left to personal belief.

From a technical standpoint, we know the living brain emits carrier waves like a radio station, which can be received by medical equipment and are generally considered a sign of life in living beings. If these waves stay anywhere near the same frequency when we pass into the spirit realm, then this leaves the definite possibility that a living person can communicate with someone who has passed over to the spirit realm, if they can emit and receive on the same wavelength. Someday we may be able to prove or disprove this scientifically without a shadow of doubt, but for now we are left with speculation.

Controversy aside and assuming direct human-to-spirit communication is possible, sensitives can offer information our electronics gadgets cannot. When both are used during an investigation, circumstantial evidence can get closer to valid phenomenon or you may find the whole haunting could be bunk, just based on what the sensitive may come up with. It's not uncommon for a possible haunted location to be the product of a fabricated story which got blown out of proportion as it was handed down.

Finding a genuine sensitive you can trust is not always easy. I've worked with many sensitives over the past couple of years, some of whom I trust implicitly and others I suspect are not what they claim to be. If they are spinning tales of floating objects, demon attacks and other bizarre, absolutely unbelievable stories, chances are that's what they are: stories. Typically, when asked for evidence of their experiences, they do not have anything to back themselves up.

Several years ago, I joined the Spirit Society of Pennsylvania, a group that was started in 1996 by psychic medium Kelly Weaver and her husband, John, as a place for people to share their paranormal experiences and learn from one another. Kelly often touts the group as an AA for the paranormal and is one of the first true mediums I have ever met.

Working with Kelly on an investigation is nothing thrilling; no theatrics, no possessions and best of all, she will tell it like it is. Though it may seem boring at times, she is not into sugar coating her experiences and making them seem bigger than they really are or making up a story just for show. She has no problem admitting when she is wrong and she doesn't scramble to cover up her mistaken impressions. In my honest opinion, she is the real deal.

There are many ways to utilize talents like Kelly's, but I find the best method to solidify their experience and your own is to send the psychic into the location with little or no information about what is happening and let them bring the information to you for verification after they have done their initial walkthrough.

Keep in mind that sensitives are people too and their impressions can be affected by their own state of mind and the attitudes of the rest of the group. Let them do what they need to do to prepare for a reading, regardless of how bizarre or ridiculous it may seem. If they are having a bad night, it may be a good idea to reschedule the walkthrough. If someone else in the group is having a bad night, it may be best for them to wait outside or somewhere else while the reading is taking place. Negative energy and unpleasant emotions just breed more negativity, which could influence what the sensitive is feeling.

If you accompany the sensitive on the walkthrough, keep an open and clear mind along with a closed mouth. Let them say what they need to and don't give them any feedback, positive or negative, until they are done. Although this may sound harsh, giving any feedback one way or another could bias their next impression. There are times when a sensitive appears to be wrong based on the information known about the location where new information is uncovered later that confirms their impressions.

Even if you are skeptical of sensitives or just downright don't believe that their claims are possible, give them a chance and you may find that their information can be just as valuable to an investigation as capturing a full-body apparition.

21. INTO THE FUTURE

Thomas Alva Edison with his phonograph

As new devices are introduced into the paranormal community, it's going to take some experimentation and button pushing to figure out how they may be used in the field. Take time to familiarize yourself with any new device. Learn as much as you can about how it operates and run it through tests at home before taking out on an investigation. The more you learn, the easier it will be to discriminate between plausible evidence and device quirks.

An open and skeptical mind is your best asset when considering possible evidence from any device regardless whether it store-bought or custom-made. There is currently no device on the market that has been proven to detect or record the existence of spirits or paranormal phenomenon; there are only theories.

Looking toward the future of paranormal technology, many speculate

Tesla Mind Viewer

we are getting closer to scientifically proving that life continues after the body gives up the ghost (pun intended.) Although it appears we have broken some ground since the days of Edison and his would-be spirit telephone, we still seem to be plagued with the same unanswered questions. Maybe you or someone you know will be instrumental in providing proof. Until then, happy hunting!

GLOSSARY

Afterlife
A continued life or existence believed to follow bodily death.

Apparition
The visual appearance of a spirit.

CCD
Charge Coupled Device. A tiny matrix of light sensors used to capture images typically found in digital cameras and most video cameras.

DAT
Digital Audio Tape. Audio tape used to capture binary data.

DRV
Direct Radio Voice. Spirit voices being heard or recorded through any kind of radio device.

Electro-Magnetic Field
EMF. An electrical disturbance which radiates from most electronic devices possibly emitted by spirits.

Electromagnetic Spectrum
The range of all known electromagnetic radiation.

EVP
Electronic Voice Phenomenon. Spirit voices captured on any type of audio recording medium.

Frank's Box
A modified AM radio designed by Frank Sumption and used to communicate with spirits.

Frequency
The amount of times an energy pattern cycles from low to high and then low again in a given amount of time.

Frequency Counter
Device used to visually show the frequency of an electrical field.

Gauss
The mathematical unit used to measure the strength of an electromagnetic field.

Ghost
A form of apparition, usually the visual appearance of a deceased human.

Gramophone
A device created by Emile Berliner around 1887 to record and playback sound on plastic or carbon discs. Invented about the same time as Edison's phonograph.

Kinetic Energy
The work needed to accelerate a body of a given mass from rest to its current velocity

LCD
Liquid Crystal Display. A small electronic device typically used to display the image output of camcorders and digital cameras.

Magnetometer
EMF detector, Gaussmeter. A device to measure the presence of a magnetic field as well as its strength, direction, and fluctuation.

Medium
A person who can communicate with spirits.

Paranormal
Above or outside the natural order of things as currently understood.

Orb
A ball of light captured in a photograph, which is speculated to be a spirit but is most likely dust illuminated by the camera's flash.

Orthicon Tube
A video device that uses a metal plate bombarded with electrons to

capture images. This device was used in the first televisions cameras up until the late 1970s.

Phonograph
A device created by Thomas Edison around 1877 that can record and play back sound on a carbon, wax or tinfoil cylinder.

Phonoautograph
A device created by Leon Scott around 1857 that shows a graphic representation of sound waves.

Psychical Research
Term coined in the late 19th century to refer to the scientific study of the paranormal.

Psychophone
Or Psychofon. A broadband radio spirit communication device based on the diode method, which was designed by Viennese engineer Franz Seidl

Radiation
Energy emitted by an object, spirit or living person that can be read by electronic instruments and felt by a sensitive.

Radio voice phenomenon
RVP. Receiving the voice of a spirit over a regular radio.

Sensitive
Someone who can sense a spirit's presence.

Shaman
A person in tribal societies who is an intermediary between the living, the dead, and the gods.

Skeptic
A person with doubts about the existence of paranormal activity.

Skeptical Mind
Using raw logic to explain an experience that would otherwise be considered paranormal.

Slave Flash
An External add-on flash which fires at the same time as the onboard camera flash to extend its range and make for a brighter scene.

Spiricom
A two-way communication device between Earth and other planes of existence created by George W. Meek in the early 1970s.

Spiritualism
A belief system that spirits of the dead can communicate with living humans in the material world. This is usually achieved through a intermediary known as a medium.

White Noise
A hiss-like sound created by tuning to a blank station on a radio or using a small electronic device to recreate a hissing sound.

Bibliography

The Ghost Hunter's Guidebook / Troy Taylor (2001)

The Encyclopedia of Ghosts and Spirits / Rosemary Ellen Guiley (2007)

Mind Over Matter / Loyd Auerbach (1996)

Field Guide to Spirit Photography / Dale Kaczmarek (2002)

American Science and Invention / Mitchell Wilson (1954)

Spiricom – An Electromagnetic-Etheric Systems Approach to Communication / George W. Meek (1982)

FATE Magazine
December 1993
July 1970
June 1974
March 1966
September 1969

www.worlditc.org

www.prairieghosts.com

ACKNOWLEDGMENTS

I would like to thank Troy Taylor for giving me the opportunity and inspiration to share my experience, experiments and knowledge with the rest of the paranormal community through the writing of this book. Troy has been helping individuals and groups throughout the country with his own books and conferences for over 15 years. I would also like to thank Rosemary Ellen Guiley who has written the forward and been an excellent source of information and inspiration. Rosemary has also been helping individuals and group with her own work. She is not dubbed Queen of the Paranormal for nothing. You wont find a more knowledgeable individual on everything paranormal.

I would also like to thank my lovely and talented wife Melissa who spent many hours reading, editing and correcting my work. Thank you for putting up with my late night experimentation on the kitchen table. Luckily I didnt burn the house down. Melissa, I love you!

In addition I am also grateful for the help and support of many fellow researches and investigators, including John and Kelly Weaver for their inspiration, knowledge, support and friendship; Rick Fisher for his guidance, friendship, support and for providing a location for weekly experimentation and the Spirit Society of Pennsylvania for offering their support and feedback.

ABOUT THE AUTHOR

Craig Telesha has been interested in technology and its internal working almost all his life. From an early age he has been dismantling and learning about every electronic device that crossed his path and earned a local reputation as "the guy who can fix just about anything."

In his early teens, he became interested in paranormal phenomenon after reading various books based on psychokinetic energy and psychic awareness. While attending a paranormal conference in early 2000 and discovering a need for more technical information to be made available to the paranormal community, he combined these two interests and started the web site www.ghost-tech.com with the mindset of demystifying the technology behind the devices used for paranormal investigations.

Since then Telesha has been using his talents to modify everyday devices for use in paranormal investigations and even designing a few of his own. He currently holds the title as Technical Director for the Spirit Society of Pennsylvania, a local paranormal group based in New Cumberland, PA which is headed up by founders Kelly and John Weaver.

ABOUT WHITECHAPEL PRESS

Whitechapel Productions Press is a division of Dark Haven Entertainment and a small press publisher, specializing in books about ghosts and hauntings. Since 1993, the company has been one of America's leading publishers of supernatural books and has produced such best-selling titles as Haunted Illinois, The Ghost Hunters Guidebook, Ghosts on Film, Confessions of a Ghost Hunter, Resurrection Mary, Bloody Chicago, The Haunting of America, Spirits of the Civil War and many others.

With nearly a dozen different authors producing high quality books on all aspects of ghosts, hauntings and the paranormal, Whitechapel Press has made its mark with America's ghost enthusiasts.

Whitechapel Press is also the publisher of the acclaimed **Ghosts of the Prairie** magazine, which started in 1997 as one of the only ghost-related magazines on the market. It continues today as a travel guide to the weird, haunted and unusual in Illinois. Each issue also includes a print version of the Whitechapel Press ghost book catalog.

You can visit Whitechapel Productions Press online and browse through our selection of ghostly titles, plus get information on ghosts and hauntings, haunted history, spirit photographs, information on ghost hunting and much more. by visiting the internet website at:

www.dark haven entertainment.com

Or call us toll-free at 1-888-446-7859 to order any of our titles.
Discounts are available to retail outlets and online booksellers!

CPSIA information can be obtained at www.ICGtesting.com
Printed in the USA
LVOW011111101111
254337LV00003B/317/P